Lecture Notes in Computer Science 16630

Founding Editors

Gerhard Goos
Juris Hartmanis

Fan Chung Graham · Megan Dewar ·
Bogumił Kamiński · Paweł Prałat ·
François Théberge

Editors

Modelling and Mining Networks

21st International Workshop, WAW 2026
Toronto, ON, Canada, June 15–19, 2026
Proceedings

Springer

Editors
Fan Chung Graham
University of California San Diego
La Jolla, CA, USA

Bogumił Kamiński
SGH Warsaw School of Economics
Warsaw, Poland

François Théberge
Tutte Institute for Mathematics
and Computing
Ottawa, ON, Canada

Megan Dewar
Tutte Institute for Mathematics
and Computing
Ottawa, ON, Canada

Paweł Prałat
Toronto Metropolitan University
Toronto, ON, Canada

ISSN 0302-9743   ISSN 1611-3349 (electronic)
Lecture Notes in Computer Science
ISBN 978-3-032-27192-1   ISBN 978-3-032-27193-8 (eBook)
https://doi.org/10.1007/978-3-032-27193-8

Preface

The *21st Workshop on Modelling and Mining Networks (WAW 2026)* was held at the Fields Institute for Research for Mathematical Sciences, Toronto, ON, Canada (June 15–19, 2026). This is an annual meeting, providing opportunities for researchers in network science to interact and to exchange research ideas. We do hope that the event was an effective venue for the dissemination of new results and for fostering research collaboration.

Virtually every human-technology interaction, or sensor network, generates observations that are in some relation with each other. As a result, many data science problems can be viewed as a study of some properties of complex networks in which nodes represent the entities that are being studied and edges represent relations between these entities. Such networks are often large-scale and decentralized, and evolve dynamically over time. Modelling and mining complex networks in order to understand the principles governing the organization and the behaviour of such networks is crucial for a broad range of fields of study, including information and social sciences, economics, biology, and neuroscience.

The aim of the *21st Workshop on Modelling and Mining Networks (WAW 2026)* was to further the understanding of networks that arise in theoretical as well as applied domains. The goal was also to stimulate the development of high-performance and scalable algorithms that exploit these networks. The workshop welcomed researchers who are working on graph-theoretic and algorithmic aspects of networks represented as graphs or hypergraphs and other higher-order structures.

This volume contains the papers accepted and presented during the workshop. Each submission was carefully reviewed in a single-blind process by four members of the Programme Committee. Papers were submitted and reviewed using the EasyChair online system. The committee members decided to accept 7 papers. Additionally, there were talks not associated with proceedings papers and 5 invited talks given by leading experts in modelling and mining complex networks: Tina Eliassi-Rad (Northeastern University), Przemysław Kazienko (Wrocław University of Science and Technology), Renaud Lambiotte (University of Oxford), Cristopher Moore (Santa Fe Institute), and Clara Stegehuis (University of Twente).

June 2026

Fan Chung Graham
Megan Dewar
Bogumił Kamiński
François Théberge
Paweł Prałat
Program Chairs

Organization

General Chairs

Andrei Z. Broder — Google Research, USA
Fan Chung Graham — University of California San Diego, USA
Paweł Prałat — Toronto Metropolitan University, Canada

Organizing Committee

Fan Chung Graham — University of California San Diego, USA
Megan Dewar — Tutte Institute for Mathematics and Computing, Canada
Bogumił Kamiński — SGH Warsaw School of Economics, Poland
Paweł Prałat — Toronto Metropolitan University, Canada
François Théberge — Tutte Institute for Mathematics and Computing, Canada

Sponsoring Institutions

Fields Institute for Research in Mathematical Sciences, Canada
NAWA – The Polish National Agency for Academic Exchange, Poland
Google, USA
SGH Warsaw School of Economics, Poland
Toronto Metropolitan University, Canada
Tutte Institute for Mathematics and Computing, Canada

Program Committee

Konstantin Avratchenkov — INRIA,France
Leman Akoglu — Carnegie Mellon University,USA
Mindaugas Bloznelis — Vilnius University,Lithuania
Paolo Boldi — University of Milan, Italy
Anthony Bonato — Toronto Metropolitan University, Canada
Ulrik Brandes — ETH Zürich, Switzerland
Piotr Bródka — Wrocław University of Science and Technology, Poland

Marian Boguña   Universitat de Barcelona, Spain
Hocine Cherifi   University of Burgundy, France
Fan Chung Graham   University of California San Diego, USA
Colin Cooper   King's College London,UK
Megan Dewar   Tutte Institute for Mathematics and
                              Computing,Canada

Andrzej Dudek   Western Michigan University, USA
Alan Frieze   Carnegie Mellon University, USA
Jeannette Janssen   Dalhousie University,Canada
Cliff Joslyn   Pacific Northwest National Laboratory, USA
Bogumił Kamiński   SGH Warsaw School of Economics, Poland
Fariba Karimi   Graz University of Technology, Austria
Julia Komjathy   Delft University of Technology, The Netherlands
Ravi Kumar   Google, USA
Nicolas Landry   University of Virginia, USA
Marc Lelarge   INRIA, France
Lasse Leskela   Aalto University, Finland
Nelly Litvak   Eindhoven University of Technology, The
                              Netherlands

Oliver Mason   Maynooth University, Ireland
Paweł Misiorek   Poznań University of Technology, Poland
Dieter Mitsche   Pontificia Universidad Católica de Chile, Chile
Peter Morters   University of Cologne, Germany
Tobias Müller   Groningen University, The Netherlands
Mariana Olvera-Cravioto   University of North Carolina at Chapel Hill, USA
Pan Peng   University of Science and Technology of China,
                              China

Xavier Perez-Gimenez   University of Nebraska-Lincoln, USA
Mason Porter   University of California Los Angeles, USA
Paweł Prałat   Toronto Metropolitan University, Canada
Katarzyna Rybarczyk   Adam Mickiewicz University, Poland
Fiona Skerman   Uppsala University, Sweden
Aaron Smith   University of Ottawa, Canada
Clara Stegehuis   Twente Universty, The Netherlands
Przemysław Szufel   SGH Warsaw School of Economics, Poland
François Théberge   Tutte Institute for Mathematics and Computing,
                              Canada

Remco van der Hofstad   Eindhoven University of Technology, The
                              Netherlands
Yana Volkovich   Microsoft, USA
Stephen Young   Pacific Northwest National Laboratory, USA
Yue Zhao   University of Southern California, USA

From Signs to Matrices on Edges: Generalising Structural Balance on Networks

Yu Tian[1] and Renaud Lambiotte[2]

[1] Center for Systems Biology Dresden, Germany
yu.tian.research@gmail.com
[2] Mathematical Institute, University of Oxford, UK
lambiotte@maths.ox.ac.uk

Abstract. Structural balance is a classical notion in network science, originally introduced in social psychology to describe globally consistent patterns of positive and negative relations. Here we present an overview of a series of works that progressively generalise this concept from signed graphs to richer classes of weighted networks. Starting from weighted signed networks, where each edge carries a real value with a sign, we revisit the classification into balanced, antibalanced, and strictly unbalanced regimes, and show how each is reflected in spectral properties and in the behaviour of spreading processes and other dynamics. We then extend the framework to networks whose edges are weighted by complex numbers, replacing the binary sign with a continuous phase, and further to matrix-weighted networks, where interactions act on multidimensional states through matrix couplings. In all these settings, the same fundamental principle applies: balance, or more generally *coherence*, describes whether signals propagating along different paths combine coherently or destructively interfere. This unifying viewpoint connects structural consistency to the spectrum of generalised adjacency and Laplacian matrices, and thereby to the long-term behaviour of both linear and nonlinear dynamics. We illustrate the reach of this perspective through consensus dynamics, random walks, spectral clustering, and synchronisation of higher-dimensional Kuramoto oscillators on networks.

1 Introduction

Networks have become a ubiquitous language for modelling interactions in complex systems, from biological circuits to social communities and technological infrastructures [1]. A central theme in network science is the interplay between the structure of a network and the dynamics that unfold on it: certain topological patterns can accelerate or suppress diffusion, shape the outcome of opinion formation, and determine whether coupled oscillators synchronise [2, 3]. Much of this understanding has been developed for networks with positive, scalar edge weights, but many real-world interactions carry richer information. In social networks, relationships can be friendly or hostile; in quantum systems, transition amplitudes are complex-valued; in image processing, the alignment between neighbouring patches is captured by an optimal transformation matrix.

Structural balance, introduced by Heider [4] and formalised by Cartwright and Harary [5], offers a powerful lens for signed networks. A signed graph is structurally balanced when every cycle contains an even number of negative edges, which is equivalent to the existence of a bipartition such that negative edges only appear between the two groups and positive edges within them. This elegant characterisation has driven research in social network analysis, opinion dynamics, biological circuits, and beyond [6, 7, 8].

The purpose of this article is to survey a series of recent results [9, 10, 11, 12] that progressively extend the notion of structural balance from the classical case of signed networks to edges with complex phases and then to the case of orthogonal matrix transformations, uncovering a unifying principle at each level: *structural balance describes whether signals propagating along different paths between two nodes interfere constructively or destructively.* This viewpoint naturally connects the structural concept of balance with spectral theory and with the asymptotic behaviour of dynamical processes on networks. In what follows, we trace this progression through three generalisations and illustrate the consequences for synchronisation of higher-dimensional oscillators.

2 Spreading and Structural Balance on Signed Networks

Consider a connected, undirected signed network $G = (V, E, \mathbf{W})$ with n nodes, where each edge weight W_{ij} is a nonzero real number whose sign encodes the nature of the interaction. The degree of a node is $d_i = \sum_j |W_{ij}|$, and the signed Laplacian is $\mathbf{L} = \mathbf{D} - \mathbf{W}$, where $\mathbf{D} = \mathrm{diag}(d_i)$. We denote by $\overline{\mathbf{W}}$ the matrix obtained by taking absolute values of all entries.

Following the classical definitions, G is *structurally balanced* if every cycle has an even number of negative edges, *structurally antibalanced* if this property holds after flipping all signs, and *strictly unbalanced* otherwise. This classification has a clean spectral counterpart. Let $\mathbf{S}$ be the diagonal matrix with entries ± 1 coming from the bipartition of the signed graph characterising the balance or antibalance structure. Then:

- If G is structurally balanced, $\mathbf{W} = \mathbf{S} \, \overline{\mathbf{W}} \, \mathbf{S}$. Therefore, $\mathbf{W}$ and $\overline{\mathbf{W}}$ share the same eigenvalues.

- If G is structurally antibalanced, $\mathbf{W} = -\mathbf{S} \, \overline{\mathbf{W}} \, \mathbf{S}$. Therefore, the spectrum of $\mathbf{W}$ negates that of $\overline{\mathbf{W}}$.

- If G is strictly unbalanced, the spectral radius strictly contracts: $\rho(\mathbf{W}) < \rho(\overline{\mathbf{W}})$.

The last point is the most revealing. The contraction arises because strictly unbalanced networks contain pairs of walks between some nodes but with opposite signs, so that the contributions from one node to another partially cancel—a form of destructive interference. This cancellation is both necessary and sufficient for strict unbalance [9].

These different spectral behaviours directly shape the behaviour of dynamics. For a linear spreading process governed by the signed adjacency matrix, balanced networks support stationary states (consensus within each group of the bipartition), antibalanced ones produce an oscillation between two configurations, and strictly unbalanced networks drive all activity to zero. A similar trichotomy appears for a nonlinear threshold model, confirming that the classification captures fundamental differences in how information propagates on signed networks [9].

3 Structural Balance with Complex Weights

Signed graphs are the special case where each edge phase is restricted to 0 or π. A natural generalisation is to allow the phase φ_{ij} to take any value in $[0, 2\pi)$, so that the weight matrix has entries $W_{ij} = r_{ij}e^{\varphi_{ij}}$ and is Hermitian [10]. Such complex-weighted networks arise in quantum mechanics (off-diagonal Hamiltonian terms), in machine learning (complex-weighted artificial neural networks), and in the study of directed networks via the magnetic Laplacian.

Structural balance is now defined through cycle phases: a connected network is *structurally balanced* if the sum of edge phases around every cycle is zero (modulo 2π), *structurally antibalanced* if this holds after shifting each edge phase by π, and *strictly unbalanced* otherwise. These definitions recover the signed case when phases are restricted to $\{0, \pi\}$. Furthermore, in complex-weighted networks, balance corresponds to the existence of a phase-consistent partition, extending the two-block sign-consistent partition that characterises balance in signed networks. The case of antibalance is similar.

The spectral results of the previous section generalise in a direct way. The role of $\mathbf{S}$ is played by a unitary diagonal matrix $\mathbf{I}_1$ encoding the phases accumulated along paths from a reference block, with $\mathbf{W} = \mathbf{I}_1^* \, \overline{\mathbf{W}} \, \mathbf{I}_1$ in the structurally balanced case, and $\mathbf{W} = -\mathbf{I}_1^* \, \overline{\mathbf{W}} \, \mathbf{I}_1$ in the structurally antibalanced case. As before, the spectral radius contracts if and only if the network is strictly unbalanced, because walks of different phases between the same endpoints produce destructive interference.

Random walks acquire a richer interpretation in this setting. A population of walkers, each carrying a phase, propagates through the network; traversing an edge adds that edge's phase to the walker's own. The averaged phase signal on each node evolves according to the complex transition matrix $\mathbf{P} = \mathbf{D}^{-1}\mathbf{W}$. If the network is structurally balanced, the process converges to a steady state where nodes in each block share the same phase—a form of local consensus. If the network is structurally antibalanced, the process exhibits oscillatory behaviour asymptotically. If the network is strictly unbalanced, the averaged phase decays to zero at every node: the diversity of phases generated by incoherent cycles washes out any preferred direction [10].

This framework also yields a spectral clustering algorithm for complex-weighted networks, generalising the notion of graph cut to accommodate phases, and provides new spectral characterisations of the magnetic Laplacian used to represent directed graphs [10].

4 Matrix-Weighted Networks and Coherence

The broadest generalisation replaces scalar edge values by matrices. In a *matrix-weighted network* (MWN), each node carries a state vector $\mathbf{y}_i \in \mathbb{R}^{n_d}$ and each edge is equipped with $\mathbf{W}_{ij} = w_{ij}\mathbf{R}_{ij}$, where $w_{ij} \geq 0$ is a magnitude and $\mathbf{R}_{ij}$ is an orthogonal matrix describing how the signal is transformed as it crosses the edge [11]. Under the reciprocity assumption $\mathbf{W}_{ij} = \mathbf{W}_{ji}^{\top}$, the supra-weight matrix $\mathcal{W}$ (with block entries $\mathbf{W}_{ij}$) is symmetric and the supra-Laplacian $\mathcal{L} = \mathcal{D} - \mathcal{W}$ is positive semi-definite, where the degree of each node is represented as $d_i\mathbf{I}$, $d_i = \sum_j w_{ij}$ and $\mathbf{I} \in \mathbb{R}^{n_d \times n_d}$. This formulation thus extends complex-weighted networks, where the effect of an edge is a rotation in two dimensions, to higher dimensions and more general linear transformations.

Structural balance becomes *coherence*: the product of transformation matrices around any directed cycle is the identity,

$$\mathbf{R}(C) = \mathbf{R}(e_1)\mathbf{R}(e_2)\cdots\mathbf{R}(e_k) = \mathbf{I}. \tag{1}$$

A coherent MWN possesses a characteristic block structure: nodes can be partitioned into groups such that intra-group transformations are trivial, inter-group transformations are consistent, and the super-graph of groups is itself coherent. Encoding this block structure in a block-diagonal matrix $\mathcal{S}$ reduces the supra-Laplacian to the scalar case:

$$\mathcal{L} = \mathcal{S}^{\top}(\bar{\mathbf{L}} \otimes \mathbf{I}_{n_d})\mathcal{S}, \tag{2}$$

where $\bar{\mathbf{L}}$ is the ordinary graph Laplacian of the magnitude-only network. The supra-Laplacian has eigenvalue zero if the MWN is coherent, and the associated eigenvectors span the subspace $\mathcal{S}^{\top}(1 \otimes \mathbf{I}_{n_d})$ [11].

The dynamical implications mirror the lower-dimensional cases. In the consensus dynamics $\dot{\mathbf{y}}_i = \sum_j w_{ij}(\mathbf{R}_{ij}\mathbf{y}_j - \mathbf{y}_i)$, coherent networks converge to a multi-consensus state where nodes in different blocks settle at different orientations related by the block transformations, while incoherent networks drive all states to zero. Random walks exhibit analogous behaviour: in coherent networks, the average state vector on each node converges to a direction dictated by the block structure; in incoherent networks, the diversity of transformations generated by conflicting cycles erases any preferred direction [11]. The level of incoherence controls the relaxation time: weakly incoherent networks first approach the multi-consensus configuration before slowly drifting toward the origin, producing a characteristic two-timescale dynamics.

5 Application: Synchronisation of Higher-Dimensional Kuramoto Oscillators

The framework of matrix-weighted networks finds a natural application in the study of synchronisation. The d-dimensional Kuramoto model describes oscillators as unit vectors $\vec{x}_i \in \mathbb{S}^{d-1}$ evolving on the sphere according to

$$\dot{\vec{x}}_i = \Omega\vec{x}_i + Z_i(\vec{x}) - \langle Z_i(\vec{x}), \vec{x}_i\rangle\vec{x}_i, \tag{3}$$

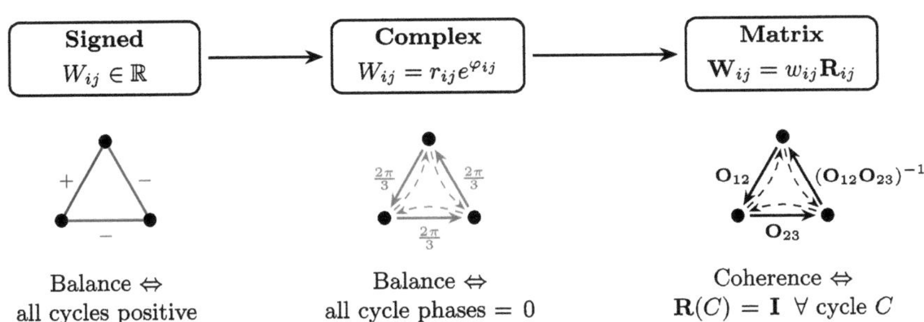

Fig. 1. The hierarchy of generalisations of structural balance (Top). At each level, an edge carries richer algebraic structure—a sign, a phase, or an orthogonal matrix—and balance (coherence) demands that the decoration accumulated around any cycle is trivial. A small triangle illustrates a balanced configuration at each level (Bottom). For the triangle with complex weights or matrix weights, only the phases of solid edges are shown, and the dashed edges of the opposite directions have the corresponding phases or transformation matrices such that the complex weight matrix is Hermitian (i.e., $\varphi_{ij} = 4\pi/3$), or the supra-weight matrix is symmetric (e.g., $\mathbf{R}_{ij} = \mathbf{O}_{12}^T$ if $\mathbf{R}_{ji} = \mathbf{O}_{12}$).

where Ω is a common antisymmetric frequency matrix and Z_i is a local mean field. On a standard scalar-weighted network, $Z_i = (K/N)\sum_j A_{ij}\vec{x}_j$, and the coupling simply averages neighbouring states. In [12], this model is extended to MWN couplings by setting $Z_i = (K/N)\sum_j w_{ij}\mathbf{R}_{ij}\vec{x}_j$, so that each edge not only transmits a signal but rotates it through the transformation $\mathbf{R}_{ij}$.

Coherence is then a necessary ingredient for synchronisation. Under the coherence condition and the additional requirement that Ω commutes with all block transformations, the change of variables $\vec{y}_i = \mathbf{O}_{1i}\vec{x}_i$ removes the rotation matrices entirely, mapping the model on a MWN to one on a scalar weighted network. This allows a Master Stability Function analysis: the stability of the synchronous solution reduces to a family of d-dimensional eigenvalue problems, each parametrised by an eigenvalue $\Lambda^{(\alpha)}$ of the scalar Laplacian $\bar{\mathbf{L}}$. Because Ω is antisymmetric, its eigenvalues are purely imaginary, and the real part of the perturbation growth rate around the synchronous state is $-K\Lambda^{(\alpha)}/N < 0$ for every non-trivial Laplacian mode. The synchronous solution is therefore locally stable for any positive coupling strength on any connected coherent MWN [12].

This result illustrates how the structural concept of coherence directly governs a genuinely nonlinear phenomenon: oscillators can lock into a common trajectory only if the network is organised so that the transformations along different paths are consistent.

6 Outlook

The progression from signed graphs ($\mathbb{Z}_2$) through complex-weighted networks ($U(1)$) to matrix-weighted networks ($O(n_d)$) reveals a unifying algebraic principle: structural balance is the condition that the composition of the group elements along any closed walk is the identity [13]. When this condition holds, a linear transformation reduces the system to a standard positive-weighted network, preserving the full spectral information. When it fails, walks carrying conflicting information interfere destructively, contracting the spectral radius and driving dynamical processes toward global consensus—a trivial zero state. Importantly, it is not the presence or the absence of edges that determines the coherence of a graph, but how the weights are assigned to existing edges.

Several avenues remain open. On the structural side, relaxed notions of coherence deserve further study: for instance, requiring only that a particular subspace, rather than the entire state space, remains invariant under cycle transformations. On the dynamical side, the interplay between coherence and nonlinearity is only beginning to be explored: beyond Kuramoto oscillators, reaction–diffusion systems and epidemic models on MWNs promise qualitatively new phenomena. Finally, on the applied side, MWNs provide a fresh perspective on multilayer networks, temporal networks, and any systems where interactions mix multiple dimensions, with random-walk–based tools (centrality, community detection) requiring extensions to this richer setting.

References

1. Newman, M.E.J.: Networks. Oxford University Press, 2nd edn. (2018)
2. Porter, M.A., Gleeson., J. P.: Dynamical Systems on Networks. Springer Cham (2016). https://doi.org/10.1007/978-3-319-26641-1
3. Lambiotte, R., Rosvall, M., Scholtes, I.: From networks to optimal higher-order models of complex systems. Nat Phys. **15**, 313–320, (2019)
4. Heider, F.: Attitudes and cognitive organization. J Psychol. **21**(1), 107–112 (1946).
5. Cartwright, D. Harary. F.: Structural balance: a generalization of Heider's theory. Psychol. Rev. **63**(5), 277–293 (1956)
6. Easley, D. Kleinberg, J.: Networks, Crowds, and Markets. Cambridge University Press (2010)
7. Facchetti, G., Iacono, G., Altafini, C.: Computing global structural balance in large-scale signed social networks. Proc. Nat. Acad.Sci. **108**(52), 20953–20958 (2011)
8. Kunegis, J., Lommatzsch, A., Bauckhage, C.: The slashdot zoo: mining a social network with negative edges. In: Proceedings of the 18th International Conference on the World Wide Web, pp. 741–750 (2009)
9. Tian, Y., Lambiotte, R.: Spreading and structural balance on signed networks. SIAM J. Appl.Math. **84**(4), 1378–1404 (2024)
10. Tian, Y., Lambiotte, R.: Structural balance and random walks on complex networks with complex weights. SIAM J. Math. Data Sci. **6**(2), 372–399 (2024)

11. Tian, S., Kojaku, Sayama, H., Lambiotte, R.: Matrix-weighted networks for modeling multidimensional dynamics: theoretical foundations and applications to network coherence. Phys. Rev.Lett. **134**, 237401 (2025)
12. Gallo, A., Lambiotte, R., Carletti, T.: Synchronization of higher-dimensional Kuramoto oscillators on networks: from scalar to matrix-weighted couplings. arXiv: 2603.08352. (2026)
13. Evans, T.S.: From signed networks to group graphs. arXiv:2505.22802 (2025)

Homophily Within and Across Groups (Based on [4])

Clara Stegehuis [iD], Abbas K. Rizi, Riccardo Michielan, Clara Stegehuis, and Mikko Kivelä

Abstract. Homophily is the tendency for nodes that share a categorical attribute to connect more often than expected under random mixing. In data, interactions occur at different group sizes, and the strength of homophily can change with the interaction size, so compressing everything into a single assortativity parameter h can miss structural trends. We model this scale dependence by generating *clique layers* G_c (one per interaction size c) on a common vertex set, assigning each layer its own homophily h_c and decomposing the global index as $h = \sum_{c \geq 2} \alpha_c h_c$. For each c, the distribution of clique types (numbers of red/blue nodes) is chosen by a maximum-entropy family that depends only on the color share and h_c, enabling straightforward likelihood-based fitting from observed clique compositions. Colored bond percolation analysis shows that redistributing the *same* global homophily across different interaction sizes can raise or lower the percolation threshold.

1 Homophily at Different Layers

Homophily is the tendency for nodes that share a categorical attribute to connect more often than expected under random mixing [1, 2]. In real data, interactions happen in different group sizes, and the sign/strength of homophily can change with in the interaction size. Collapsing all sizes to a single homophily parameter h can therefore miss important trends in the sata. Our clique-layer model assigns a homophily parameter h_c to each interaction size to model different interaction styles in different group sizes.

Model. We generate a model that captures homophily at different interaction sizes. We build G by superposing *clique layers* G_c for sizes $c \in \{2, 3, \ldots\}$ on the same set of N vertices. Each G_c only contains c-cliques, and the final graph $G = \cup_c G_c$. Next to this, each of the N nodes are colored red or blue, with n_r denoting the fraction of red notes, and $n_b = 1 - n_r$ the fraction of blue nodes. Each clique of size c can contain different numbers of red and blue nodes, and we say that a c-clique is *type-i* if it contains i red and $c - i$ blue nodes.

To generate G_c, we sample $M_c = O(N)$ cliques independently. Each clique is assigned a type, where the probability that a sampled c-clique is type-i is given by $F_{c,i}$. To sample the type i cliques of size c, we sample i red vertices uniformly, and $c - i$ blue vertices, and connect these into a clique.

As long as the maximum clique size is finite and the number of sampled cliques per layer scales linearly in N, the network remains sparse. Intuitively, the layer with $c = 2$ encodes bridging behavior (the connections between members of different groups), while $c > 2$ encodes within-group behavior. When the network is formed by $c = 2$ only, the model reduces to a two-block Stochastic Blockmodel [3] with red/blue communities.

Color balance imposes

$$\sum_{i=0}^{c} \frac{i}{c} F_{c,i} = n_r, \text{ equivalently } \sum_{i=0}^{c} \frac{c-i}{c} F_{c,i} = n_b. \tag{1}$$

Clique layers and sparsity. Because cliques are sampled independently and the maximum c is finite, overlaps between cliques vanish in the large-N limit (sparsity), and the degree distribution becomes approximately Poisson centered at a prescribed $\langle d \rangle$ set by the total number of sampled cliques across layers.

Clique homophily. Let e_{kk} be the fraction of k–k edges ($k \in \{r, b\}$). The Coleman assortativity is

$$h = \frac{(e_{rr} + e_{bb}) - (n_r^2 + n_b^2)}{1 - (n_r^2 + n_b^2)} \in [-1, 1]. \tag{2}$$

In our construction, h *decomposes* over clique sizes:

$$h = \sum_{c \geq 2} \alpha_c h_c, \quad \alpha_c = \frac{c(c-1)M_c}{N \langle d \rangle}, \tag{3}$$

where α_c is the fraction of edges contributed by layer c, and h_c is the homophily of layer c. Moreover, h_c satisfies

$$h_c = \sum_{i=0}^{c} h_{c,i} F_{c,i}, \quad h_{c,i} = \frac{\dfrac{\binom{i}{2} + \binom{c-i}{2}}{\binom{c}{2}} - (n_r^2 + n_b^2)}{1 - n_r^2 - n_b^2}. \tag{4}$$

Here, the first term in the fraction of $h_{c,i}$ the fraction of same-color edges ($e_{bb} + e_{rr}$) inside a type-i clique.

2 Choosing F_c: a Maximum Entropy Approach

Many different $\{F_{c,i}\}$ achieve the same homophily. To choose an F_c distribution consistent with these sufficient statistics, we maximize entropy under the constraints (1) and

(4). The resulting exponential family per layer is

$$F_{c,i} = \frac{1}{Z_c(\theta_r, \theta_{h_c})} \exp(\theta_r i + \theta_{h_c} h_{c,i}), \tag{5}$$

where θ_r sets the color share n_r and θ_{h_c} tunes the target layer homophily h_c; Z_c a normalizing constant.

Fitting to data. Given an empirical graph with node colors, extract all maximal cliques, group by clique size c, and compute h_c. Estimate F_c by the maximum likelihood in (5) for each c separately.

3 Percolation

We study colored bond percolation [5] where an edge between colors k–j is removed with probability π_{kj}, summarized by $\pi = (\pi_{rr}, \pi_{rb}, \pi_{bb})$. We can view component growth as a multi-type branching process on *clique types*: from a currently occupied node in a type-I clique (which specifies c and its red/blue counts), how many cliques of type J are reached through that percolated clique? These expected numbers of "offspring" cliques of each type reached from each current clique type form a nonnegative matrix $B(\pi)$.

The giant-component threshold occurs when the spectral radius of $B(\pi)$ crosses 1. This also shows that two networks with identical $\langle d \rangle$ and h can exhibit markedly different percolation thresholds because their $\{h_c\}$ profiles differ.

References

1. McPherson, M., Smith-Lovin, L., Cook, J. M.: Birds of a feather: homophily in social networks. Annu. Rev. Sociol. **27**, 415–444 (2001)
2. Newman, M. E. J.: Mixing patterns in networks. Phys. Rev. **67**, 026126 (2003)
3. Holland, P. W., Laskey, K. B., Leinhardt, S.: Stochastic blockmodels: first steps. Soc. Netw. **5**, 109–137 (1983)
4. Rizi, A. K., Michielan, R., Stegehuis, C., Kivelä, M.: Homophily withon and across groups. Nat Comm. **16**, 11351 (2025)
5. Kryven, I.: Bond percolation in coloured and multiplex networks. Nat. Commun. **10**, 404 (2019)

Contents

Twinning Complex Networked Systems: Data-Driven Calibration of the mABCD Synthetic Graph Generator

Piotr Bródka[1,2], Michał Czuba[1,2(✉)], Bogumił Kamiński[3], Łukasz Kraiński[3], Katarzyna Musial[2], Paweł Prałat[4], and Mateusz Stolarski[1]

[1] Department of Artificial Intelligence, Wrocław University of Science and Technology, Wrocław, Poland
[2] Data Science Institute, University of Technology Sydney, Sydney, Australia
michal.czuba@pwr.edu.pl
[3] Decision Analysis and Support Unit, SGH Warsaw School of Economics, Warsaw, Poland
[4] Department of Mathematics, Toronto Metropolitan University, Toronto, Canada

Abstract. The increasing availability of relational data has contributed to a growing reliance on network-based representations of complex systems. Over time, these models have evolved to capture more nuanced properties, such as the heterogeneity of relationships, leading to the concept of multilayer networks. However, the analysis and evaluation of methods for these structures is often hindered by the limited availability of large-scale empirical data. As a result, graph generators are commonly used as a workaround, albeit at the cost of introducing systematic biases. In this paper, we address the inverse-generator problem by inferring the configuration parameters of a multilayer network generator, mABCD, from a real-world system. Our goal is to identify parameter settings that enable the generator to produce synthetic networks that act as digital twins of the original structure. We propose a method for estimating matching configurations and for quantifying the associated error. Our results demonstrate that this task is non-trivial, as strong interdependencies between configuration parameters weaken independent estimation and instead favour a joint-prediction approach.

Keywords: Complex Systems · Digital Twins · Inverse Graph Modelling · mABCD · Multilayer Networks

1 Introduction

Complex networked systems are typically represented by graphs constructed from relational data. This representation facilitates the encapsulation of topological characteristics, such as relationship structures, connectivity patterns, or node centralities, which are not readily apparent at first glance [9].

To accurately characterise dynamics in complex networked systems, it is necessary to adopt an approach that differentiates between the various types of

F. C. Graham et al. (Eds.): WAW 2026, LNCS 16630, pp. 1–17, 2026.
https://doi.org/10.1007/978-3-032-27193-8_1

relationships present. A natural way to achieve this is through the use of *multi-layer networks* [16], a system defined as a collection of ℓ layers, where each layer represents a distinct mode of interaction encoded by its corresponding graph. The resulting dynamics are governed not only by intra-layer edges, describing interactions within a given relationship type, but also by the presence of the system's actors across different layers, through which they serve as inter-layer bridges. This structure enables modelling of complex spreading mechanisms that are obscured in monoplex representations [7].

Understanding the topology of a system and the interplay between the relationships it contains is critical for designing effective intervention policies. For instance, in simplified models of human interactions, immunisation strategies (such as node removal or fact-checking) typically target high-degree hubs [22]. However, in multilayer systems, the most influential actors are often not the hubs within individual layers, but rather multiplex bridges, that is, nodes with moderate intra-layer connectivity but high "participation coefficients" across layers. These actors facilitate the spillover of misinformation from fringe, weakly moderated communities (e.g., *4chan*) into mainstream discourse (e.g., *X* or *Facebook*). Consequently, observing the system through a multilayer lens reveals that resilience in one layer can mask catastrophic fragility in the coupled system. A misinformation cascade that appears sub-critical (dying out) on one platform may be sustained by activity on another, creating a hysteresis loop where the false belief becomes endemic despite platform-specific moderation efforts.

While empirical analysis of real-world datasets is foundational, it faces severe limitations when dissecting the complex dynamics of multilayer systems. Real-world multilayer data is often sparse, proprietary, or plagued by missing links, and crucially, it represents only a single, static realisation of a dynamic process. One cannot "rewind" the Internet to see how a misinformation campaign would have unfolded if the user base were 50% larger or if the moderation algorithm were different. To overcome these constraints, researchers must turn to synthetic network generation, that is, creating artificial yet statistically realistic graphs that serve as controllable laboratories for simulation. The primary utility of these synthetic environments lies in their flexibility. By employing generative random graph models, one can systematically vary topological parameters to isolate their effects on dynamical processes.

This approach culminates in the concept of the *"digital twin"* [24]. Digital twins for complex networked systems provide a dynamic virtual counterpart to real-world networks, enabling continuous monitoring, simulation, and prediction of system behaviour. Their defining strength lies in real-time data exchange between the physical system and its digital representation, allowing the model to evolve as the network evolves. By integrating simulation, optimisation, data analytics, and machine learning, digital twins capture both the structure and the dynamics of interconnected systems such as social networks, cyber-physical systems, and blockchain-based architectures. This makes them uniquely suited to exploring scenarios that have never occurred, detecting anomalies, and supporting informed decision-making in complex, distributed environments. Even on the simplest level of modelling, a digital twin is not merely a random graph, but a synthetic replica carefully calibrated to match the specific statistical properties of a target real-world system (such as its degree distribution and distribution of community sizes, etc.). Therefore, the modelling paradigm they convey makes them a powerful approach for understanding and managing complex systems.

While the utility of "digital twins" is clear, constructing one that faithfully mimics a specific real-world system remains a formidable mathematical challenge. This is fundamentally an inverse problem: given an observed set of real-world statistics, one must infer the precise combination of generative parameters that produces a matching graph. In multilayer systems, this difficulty is compounded by the curse of dimensionality. A model rich enough to capture the nuance of real systems, accounting for intra-layer topology and inter-layer interdependence, inevitably possesses a vast parameter space. Developing robust, algorithmic frameworks to automatically calibrate these high-dimensional synthetic models to empirical data is not just a technical refinement; it is a necessary frontier for making network science predictive rather than merely descriptive.

In this paper, we aim to shed light on this problem by proposing a modular approach for inferring the parameters of the mABCD generator from an observed real-world network. In particular, we investigate whether decomposing the task into a sequence of analytical and optimisation sub-problems constitutes a viable strategy. Our results indicate that the proposed approach establishes a strong baseline that could be difficult to surpass. Nevertheless, the experiments suggest that jointly predicting all configuration parameters within a single optimisation procedure may yield superior performance, which delineates an interesting yet challenging new research direction.

The remainder of the paper is organised as follows. Section 2 briefly describes the fundamental concepts underpinning the work; in particular, operating principles of the mABCD model. Section 3 presents the parameter estimation framework and introduces the discrepancy measures used to quantify approximation error. Section 4 discusses the experimental results, while Sect. 5 concludes the paper.

2 Preliminaries

To address the problem studied in this paper, we first introduce the fundamental concepts. We briefly review multilayer networks, outline the general concept for the configuration inference task, and describe the synthetic graph generator employed.

2.1 Multilayer Networks

As stated in the introduction, heterogeneity of complex networked systems is a crucial property which cannot be neglected. Therefore, the problem tackled in this work aims to take into account this property by utilising multilayer networks. Following [16], we formalise the concept of a multilayer network below. Nevertheless, before we do this, we clarify that for a given $n \in \mathbb{N} = \{1, 2, \ldots\}$,

$[n]$ is used to denote the set consisting of the first n natural numbers, that is, $[n] = \{1, 2, \ldots, n\}$. Furthermore, for a matrix $\mathbf{X}$, we denote its (i, j)-th entry by $\mathbf{X}_{ij}$.

Definition 1 (Multilayer network). *A multilayer network can be described as a quadruple $G = ([n], [\ell], V, E)$, where $[n]$ is a set of n actors, $[\ell]$ is a set of layers, $V \subseteq [n] \times [\ell]$ is a set of nodes, $E \subseteq \bigcup_{i \in [\ell]} [V_i \times V_i]$ is a set of edges.*

When analysing Definition 1, one can note that it is in line with the understanding of multilayer networks as collections of graphs, where each layer $G_i = (V_i, E_i)$ is spanned on a subset of actors existing in the system (v_i) within the relationship (layer) $i \in [\ell]$. For instance, in a system comprising interactions on various social platforms, $[n]$ would denote humans active on the Internet, $[\ell]$ social platforms, V would be a set of all accounts in the systems, i.e. nodes, where node $v = (a, \ell_i) \in V$ represents an actor a in layer ℓ_i, e.g., a *LinkedIn* profile of a citizen John Smith. On the other hand, the network could be decomposed into ℓ interdependent graphs from different social systems.

2.2   Complex Networked System Analysis with Twinning Approach

Utilising the example introduced in Sect. 1, a conceptual framework guiding the problem is presented here. Suppose one is to evaluate several "what-if" scenarios that cover interventions in a social network to design effective strategies to control the spread of misinformation, given an initial dataset of ground-truth interactions on the platform.

On the grounds presented in Sect. 1, a twin-based framework can be introduced, as shown in Fig. 1. It assumes that the analysed real-world multirelational system is encoded into a configuration file for a synthetic graph generator, which can then be used for further modelling. By altering selected parameters of the generator, one can explore a variety of what-if scenarios and construct adjusted "digital twins" that reflect possible modifications of the original system. These can be evaluated under the assumed spreading regime, allowing an assessment of which topology is optimal from the perspective of a given influence control task (e.g., maximisation). For instance, one can rigorously test hypotheses by scaling the network size to observe asymptotic behaviours, increasing edge density to identify percolation thresholds, or introducing additional layers to simulate the emergence of new communication channels. This capability is essential for establishing causal links between network structure and process outcomes, something that is impossible with observational data alone.

The most important factor affecting the validity of the approach involves three types of errors. The first one arises in the *configuration retrieval* process. For models allowing rich modelling space (like mABCD), this task is non-trivial, as some parameters cannot be directly inferred from the network and need to be estimated (e.g., a community breakdown). The second type of error relates to imperfections of the synthetic graph generator, as the produced network may deviate from the specified configuration. Finally, the third type of error arises

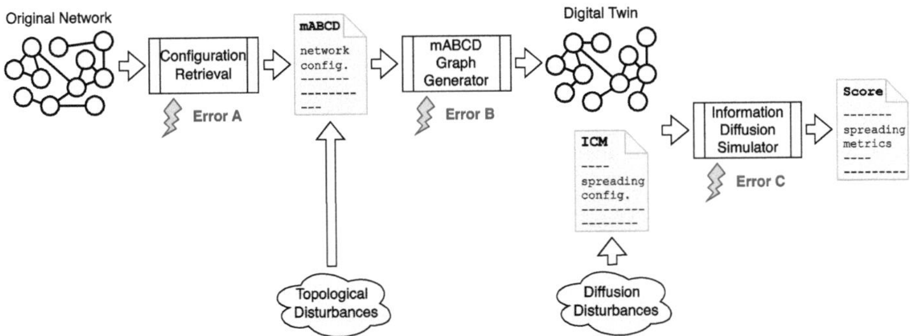

Fig. 1. Spread control pipeline utilising the configuration retrieval approach to create adjusted twins of the evaluated system. The process begins by determining the parameters of mABCD that match the analysed system. Its digital twin is then generated and evaluated under the appropriate dynamics. Note that the pipeline's efficacy depends on addressing inherent errors.

in the simulation of information diffusion, since spreading models remain only approximations of real-world processes. Addressing these errors is essential to ensure the pipeline's correctness. Given the size of the task, we aim to address the first of the presented errors in this work.

Configuration retrieval can be carried out in different ways depending on needs and availabilities. The most natural approach is to estimate each parameter independently by designing appropriate algorithms and metrics to quantify estimation quality. This, however, is not the only method. Another, perhaps complementary, natural way of measuring the "distance" between the original network G and its produced twin $\hat{G}$ would be to embed both graphs in the same high-dimensional space and use the distance between the two representations to measure the quality (*Open Problem 1*). Given the recent advancements in machine learning, this approach also feels natural, and some fast algorithms were recently introduced [8]. Nevertheless, approaching the problem without trying the more straightforward methods deprives us of a reliable baseline. Therefore, this paper tackles the problem from the former position.

2.3   The Synthetic Network Generators—mABCD

Synthetic graph generators with explicit community structure play a central role in the development and evaluation of analysis methods for complex networks, particularly in settings where large-scale empirical datasets with reliably annotated communities are scarce. To address this limitation, several benchmark models have been proposed to generate artificial networks that resemble real-world systems while retaining full control over their structural properties. A prominent example is the LFR model [18] and its extensions [5], which allows for heterogeneous degree distributions and community sizes and has become a standard tool for evaluating community detection algorithms.

6   P. Bródka et al.

More recently, the Artificial Benchmark for Community Detection (ABCD) model [13] has been introduced as a scalable [11] and theoretically tractable alternative to LFR. Undirected variants of ABCD generate networks with comparable structural properties, while offering improved computational performance and greater flexibility in interpolating between well-defined community structure and random graphs [11]. Owing to these properties, ABCD has been extensively analysed from a theoretical perspective, including studies of its modularity behaviour [12] and self-similarity properties [3]. The modular design of the model has further enabled a range of extensions, including outliers [14], overlapping communities [2], hypergraphs [15], and multilayer networks [17]. The multilayer extension, mABCD, generalises the ABCD framework to systems in which multiple interaction layers jointly shape network dynamics. Rather than aiming to exactly reproduce a specific empirical network, mABCD is intended to generate ensembles of graphs that preserve key structural characteristics of a target system, and can therefore be interpreted as coarse-grained digital twins of complex networked systems.

In this work, we adopt mABCD as the synthetic generator underlying our experimental analysis. Therefore, to focus on the aspects most relevant to the configuration inference problem considered here, a brief review of its core principles is necessary. In principle, the mABCD network generation comprises of six subsequent phases. The process is governed by global and local (independent for each layer) parameters presented briefly in Table 1. For a detailed description, please refer to [17].

Table 1. Global and local parameters of mABCD denoted together as Θ.

Parameter	Range	Description
Global parameters		
n	$\mathbb{N}$	Number of actors
ℓ	$\mathbb{N}$	Number of layers
d	$\mathbb{N}$	Dimension of the biscuit-like reference layer
$\mathbf{R}$	$[0,1]^{\ell \times \ell}$	Inter-layer edge correlation matrix
Local parameters (independent for each i layer)		
q_i	$(0,1]$	Fraction of active actors
τ_i	$[-1,1]$	Correlation coef. between degrees and actors' labels
r_i	$[0,1]$	Correlation strength between communities and the reference layer
γ_i	$(2,3)$	Power-law degree distr. with exponent γ_i
δ_i	$\mathbb{N}$	Min. degree
Δ_i	$\mathbb{N}\,(1 \leq \delta_i \leq \Delta_i < n)$	Max. degree
β_i	$(1,2)$	Power-law community size distr. with exponent β_i
s_i	$\mathbb{N}$	Min. community size
S_i	$\mathbb{N}\,(\delta < s_i \leq S_i \leq n)$	Max. community size
ξ_i	$(0,1)$	Level of noise

The first five steps are carried out independently for each of the ℓ layers, while the sixth one binds the generated graphs into a single multilayer network. In the first phase, the so-called active nodes (i.e., those that will not be isolated) are determined. Each active node is endowed with its degree in the second step. During the third step, communities (understood as in [6]) are created and populated with nodes. This is done with the help of the reference latent layer, which is implicitly generated during the network generation process and discarded afterwards. It serves as a proxy for actors' features, which typically shape connections in complex networked systems. The layer can be visualised as a d-dimensional biscuit, with raisins representing actors: the closer two raisins are, the more likely the corresponding actors are to be assigned to the same community. Phase four focuses on connecting nodes within the communities as well as establishing inter-community (i.e., background) links. These graphs are subsequently simplified to remove possible self-loops and multiedges while preserving the intended structural properties. The final phase performs a series of edge rewirings to achieve the targeted edge correlations between each pair of layers.

3   Proposed mABCD Configuration Retrieval Method

Suppose that we are given a multilayer network following Definition 1. Our goal is to generate a digital twin, a multilayer network $\hat{G}$ generated by mABCD with associated graphs $\hat{G}_i = (\hat{V}_i, \hat{E}_i)$ for $i \in [\ell]$, that mimics the original network G as best as possible. In order to do it, one needs to appropriately select the parameters Θ of mABCD and then measure the "success", i.e., how close G to $\hat{G}$ is. The success is measured by computing a *divergence score* $\mathcal{D}(\hat{G}, G)$; the smaller the score, the better the fit is obtained.

As already noted, mABCD consists of several parameters, but not all of them are used directly to model the target network's properties. Some should be treated rather like hyperparameters, by manipulating which the accuracy of transforming a given configuration setup into the final network can be controlled. These were intentionally omitted in Sect. 2.3, as they are out of our interest due to a loose connection with the structural properties of the produced network, but we acknowledge them to show the complexity of the problem. In the proposed approach, we fix them to the values recommended in [17].

The remaining mABCD parameters (Table 1) explicitly provide expected structural properties of the network and can be grouped by their proximity. Based on the observation, we propose a method that can independently predict each parameter group by leveraging an algorithmic approach. The only exemption is r, which cannot be feasibly estimated as it employs the latent layer used to model actor features—neither returned by the model nor available in real-world networks. That is why a solution-search-browsing technique was employed, i.e., the Bayesian optimisation.

Parameters such as n, ℓ, δ_i, and Δ_i are trivial to extract from G. Similarly, the fraction of active actors is easy to obtain: $q_i = |V_i|/|\bigcup_{i \in [\ell]} V_i|$.

Dimension of Reference Latent Layer (d). Communities in mABCD are independently generated in each layer, but the desired correlations are obtained via a hidden, low-dimensional geometric reference layer. Nodes close to each other in the reference layer end up in the same community with a larger probability. In particular, the dimension of a reference layer d affects many properties of the generated network (see Sec. 4 in [17]). However, it seems not so easy to guess the right value so we leave it as a parameter that eventually can be tuned for the best outcome, that is, the smallest divergence score $\mathcal{D}(\hat{G}, G)$. Understanding the influence of this hyperparameter and selecting the optimal value is left as an open problem (*Open Problem 2*).

This geometric approach to community structure is justified by the fact that latent geometric spaces are widely believed to shape complex networks (e.g., social media networks shaped by users' opinions, education, knowledge, interests, etc.). These latent spaces have been successfully employed for many years to model and explain network properties such as self-similarity [21], homophily, and aversion [10]. For further references, we direct the reader to the survey [4] or the book [20].

Inter-layer Edge Correlation $(\mathbf{R})$. To measure correlations between edges in different layers, we define $\mathbf{R}$, an $\ell \times \ell$ matrix in which elements $\mathbf{R}_{i,j} \in [0, 1]$ $(i, j \in [\ell])$ capture correlation between edges present in layers i and j. For any $i, j \in [\ell]$, let E_i^j be the set of edges that are present in layer i, involving actors that are also active in layer j. Entries in $\mathbf{R}$ are computed using the following formula:

$$\mathbf{R}_{i,j} = \frac{|E_i^j \cap E_j^i|}{\min\{|E_i^j|, |E_j^i|\}}.$$

If $\min\{|E_i^j|, |E_j^i|\} = 0$, then we leave $\mathbf{R}_{i,j}$ undefined.

Note that $\mathbf{R}_{i,i} = 1$ for any $i \in [\ell]$ and $\mathbf{R}_{i,j} = \mathbf{R}_{j,i}$ for $1 \leq i < j \leq \ell$. The maximum value of 1 is attained when edges in one of the layers form a subset of edges in the other layer. The minimum value of 0 is attained when the two sets of edges in the corresponding layers are completely disjoint.

Extracting $\mathbf{R}$ from the original network G is straightforward. The mABCD model aims to produce a synthetic network with the desired edge correlation matrix, but this is not guaranteed, and some small error always occurs. Let $\hat{\mathbf{R}}$ be the corresponding edge correlation matrix for the obtained digital twin $\hat{G}$. The divergence score is simply the normalised to $[0, 1]$ Frobenius norm between the two matrices:

$$\mathcal{D}_{\mathbf{R}}(\hat{G}, G) = \frac{\|\mathbf{R} - \hat{\mathbf{R}}\|_F}{\sqrt{\ell(\ell - 1)}} = \sqrt{\frac{1}{\ell(\ell - 1)} \sum_{i,j \in [\ell]} (\mathbf{R}_{ij} - \hat{\mathbf{R}}_{ij})^2} = \sqrt{\frac{1}{\binom{\ell}{2}} \sum_{1 \leq i < j \leq \ell} (\mathbf{R}_{ij} - \hat{\mathbf{R}}_{ij})^2}.$$

Correlation Coefficient Between Degrees and Labels (τ_i). This family of parameters models correlations between sequences of node degrees that are active in two different layers. The mABCD model assumes that the labels of actors and nodes representing them are natural numbers. The parameter $\tau_i \in [-1, 1]$ models the correlation between these labels and the corresponding degree sequence in layer i.

To estimate τ_i, which relies on the ordering of actor labels, we first sort actors with respect to their total degree (sum of degrees over all layers), breaking ties randomly if needed; that is, the actor with the highest total degree is assigned label n, the highest possible label. Then, for any $i \in [\ell]$, we compute the Kendall rank correlation coefficient τ_i between these labels and the corresponding degree sequence in layer i. Moreover, only nodes with a positive degree are taken into account to obtain the correlation value in order to reduce the noise; i.e., inactive nodes are ignored as they should be.

To measure the quality of the fit, one can compute an $\ell \times \ell$ matrix $\mathbf{A}$, where $\mathbf{A}_{ij} \in [-1, 1]$ is the Kendall rank correlation coefficient between layers i and j for the real network G. Clearly, only nodes that are active in both layers (i.e., nodes in $V_i \cap V_j$) are considered. Once a digital twin is generated, matrix $\hat{\mathbf{A}}$ is computed analogously, but this time for the twin $\hat{G}$. As before, the divergence score is simply the (normalised) Frobenius norm between the two matrices:

$$\mathcal{D}_\tau(\hat{G}, G) = \frac{\|\mathbf{A} - \hat{\mathbf{A}}\|_F}{\sqrt{4\ell(\ell-1)}} = \sqrt{\frac{1}{4\binom{\ell}{2}} \sum_{1 \leq i < j \leq \ell} (\mathbf{A}_{ij} - \hat{\mathbf{A}}_{ij})^2}.$$

Correlation Strength Between Communities and the Reference Layer (r_i). This family of parameters, which models correlations between communities formed by nodes that are active in two different layers, appears to be the most challenging to retrieve. That results from inaccessibility of the latent biscuit-like reference layer (whence the communities are sampled and whose dimension is modelled by d) used during network generation. However, once a digital twin is generated, the success of estimating r can be assessed by computing the corresponding divergence score $\mathcal{D}_r(\hat{G}, G)$, which measures a feature indirectly captured by r—the cross-layer alignment of communities between G and $\hat{G}$. By efficiently searching the parameter space to identify the value $\hat{r}$ that minimises $\mathcal{D}_r(\hat{G}, G)$, one can attempt to fit this parameter.

Let us then concentrate on the divergence score. For each layer $i \in [\ell]$ in the real network G, we independently run some stable clustering algorithm (in experiments, the Greedy Modularity Optimisation [6]) to get a partition $\mathcal{P}_i$ that identifies its community structure. Then, we compute an $\ell \times \ell$ matrix $\mathbf{B}$, where $\mathbf{B}_{ij} \in [0, 1]$ is the adjusted mutual information (AMI) between partitions $\mathcal{P}_i$ and $\mathcal{P}_j$ induced by nodes that are active in both layers, that is, partitions induced by $V_i \cap V_j$. Matrix $\hat{\mathbf{B}}$ is computed analogously but for the twin $\hat{G}$ and the divergence score is defined as follows:

$$\mathcal{D}_r(\hat{G}, G) = \frac{\|\mathbf{B} - \hat{\mathbf{B}}\|_F}{\sqrt{\ell(\ell-1)}} = \sqrt{\frac{1}{\binom{\ell}{2}} \sum_{1 \leq i < j \leq \ell} (\mathbf{B}_{ij} - \hat{\mathbf{B}}_{ij})^2}.$$

The proposed method for predicting r is based on access to the previously estimated parameters of the mABCD configuration (Θ) and the $\mathbf{B}$ matrix of the original network. The objective function generates n candidate twins from the configuration $\Theta^{(r)}$ corresponding to the evaluated value of r. It then computes a $\hat{\mathbf{B}}$ matrix for each twin and returns the divergence score $\mathcal{D}_r$ between $\mathbf{B}$ and the mean matrix obtained from the collection of $\hat{\mathbf{B}}$ matrices, denoted by $\bar{\mathbf{B}}$. This value is subsequently minimised during the Bayesian optimisation process to obtain the best-matching vector $\hat{r}$:

$$\hat{r} = \arg \min_{r \in [0,1]^\ell} \mathcal{D}_r(\mathbf{B}, \bar{\mathbf{B}}) : \quad \bar{\mathbf{B}} = \frac{1}{n} \sum_{k=1}^{n} \hat{\mathbf{B}}^{(k)}(\Theta^{(r)})$$

By sequentially evaluating the objective function at values of r selected based on previous evaluations, the method seeks to identify the best-matching value of r. The implementation employed, provided by the skopt [19] library, accounts for the non-deterministic behaviour of mABCD by incorporating Gaussian noise into the surrogate model of the objective. A typical optimisation trajectory for fitting r is shown in Fig. 2.

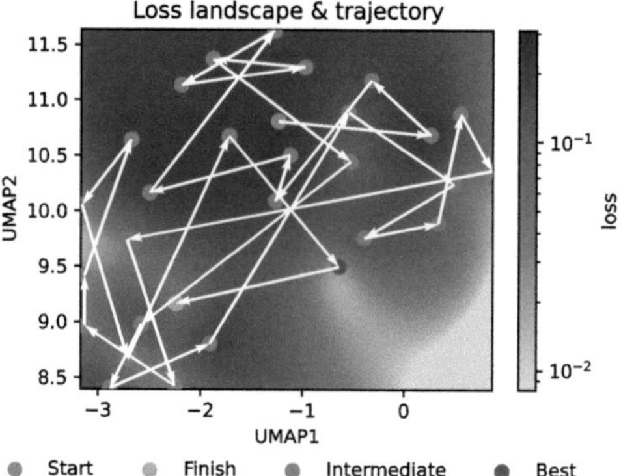

Fig. 2. A typical optimisation trajectory of r projected on $\mathbb{R}^2$ space obtained during experiments. Note the preserved trade-off between exploration and exploitation, by which the employed method avoids getting stuck in local minima.

Degree Distributions $(\gamma_i, \delta_i, \Delta_i)$. This family of parameters models the degree sequences in the corresponding layers. Among them, only the retrieval of γ_i is non-trivial, as it involves the general challenge of fitting an observed distribution to a specific model. To estimate this parameter, we employ the powerlaw library [1], implicitly assuming that the degree distribution follows a power-law.

The **mABCD** model generates, by design, power-law sequences with parameters provided as the input. Hence, the only discrepancy between the original network G and its synthetic twin $\hat{G}$ comes from fitting the power-law distribution to the observed degree sequence of the real network. To measure the quality of the fit, we employ the Kolmogorov–Smirnov test. It is a nonparametric statistical test used to determine if a sample comes from a specific distribution (one-sample) or if two samples come from the same distribution (two-sample) by measuring the maximum difference between their cumulative distribution functions. This leads to the following divergence score:

$$\mathcal{D}_\gamma(\hat{G}, G) = \frac{1}{\ell} \sum_{i \in [\ell]} \max_{k \in [\delta_i, \Delta_i]} \left| \frac{N_i(k)}{N_i(\delta_i)} - \frac{\int_k^\infty x^{-\gamma_i} dx}{\int_{\delta_i}^\infty x^{-\gamma_i} dx} \right|,$$

where $N_i(k)$ is the number of nodes of degree at least k in layer i (as a result, as usual, we ignore inactive nodes in this layer).

For simplicity, in this paper we assume that the degree sequences within layers of a real multilayered network G (i.e., specific to nodes rather than actors) approximately follow a power law. But not all networks have degree sequences of this nature. To solve this issue, the **mABCD** model is flexible and allows the degree sequences to be directly injected into the model. Since **mABCD** is using the classical configuration model to generate edges in each layer, with this more advanced option, the degree sequence of the digital twin $\hat{G}$ would match *exactly* the original network G. We leave an implementation of this extension into our framework and an investigation of its consequences on the quality of the digital twin for the future (*Open Problem 3*).

Distributions of Community Sizes (β_i, s_i, S_i). The next family of parameters governs distributions of community sizes in the corresponding layers. Recall that we already extracted the community structure in each layer $i \in [\ell]$ of the real network G by running some stable clustering algorithm, partitions $\mathcal{P}_i$. We may now compute the corresponding sequences of community sizes and extract the values of s_i and S_i from these sequences. Following the same procedure as for dealing with the degree distributions, one can estimate the parameters β_i and measure the success by computing the corresponding divergence score $\mathcal{D}_\beta(\hat{G}, G)$. As before, we leave it for future work to investigate how much one gains if the sequence of community sizes is directly injected into the **mABCD** model (*Open Problem 4*).

Level of Noise (ξ_i). Finally, to estimate the layer-wise noise level between communities, we again analyse the community structure $\mathcal{P}_i$ of each network layer $i \in [\ell]$. For each layer i of the original multilayered network G, the parameter ξ_i is extracted by simply looking at the fraction of inter-community edges (edges connecting nodes from different parts of the partition $\mathcal{P}_i$) relative to the total number of edges in the layer i. The **mABCD** model aims to preserve this property, but some very small error could be introduced. More importantly, the model preserves the level of noise between the ground-truth partition, which might not be exactly the same as the partition found by the clustering algorithm (especially for noisy graphs). Hence, once the twin $\hat{G}$ is generated, we extract the associated parameters $\hat{\xi}_i$ and compute the divergence score:

$$\mathcal{D}_\xi(\hat{G}, G) = \sqrt{\frac{1}{\ell} \sum_{i \in [\ell]} (\xi_i - \hat{\xi}_i)^2}.$$

Cumulative Divergence Score. The parameters of mABCD affect some important properties of multilayered networks; hence, it is desired for the digital twin $\hat{G}$ to match these properties as best as possible. For each parameter of the model, we introduced above some natural ways to measure how well the corresponding property is preserved via their divergence scores. To estimate an overall quality of the digital twin, one needs to somehow combine these scores. Their definitions guarantee that each of the scores is in $[0, 1]$, but they are not comparable. Hence, before we develop a framework for generating good digital twins, we need to come up with a good measure of quality that combines all aspects together (*Open Problem 5*).

4   Experiments

To test the proposed configuration retrieval method, two experiments using the framework proposed in Sect. 3 were performed. In both of them a real-world multilayer network *Freebase* [23], representing cooperation in the movie industry, was employed. The *Freebase* network captures relations between movies and people involved in their production (**Actors**, **Writers**, **Directors**). No explicit inter-layer edges are defined (which aligns with Definition 1); layers correspond to meta-path relations (**MAM**, **MDM**, **MWM**) over the same movie nodes. The dataset is used here only as an illustrative example, and its semantics are not considered further. The network has 3,492 actors; the numbers of nodes per layer are 3,479, 2,091, and 1,865, with 129,097, 7,099, and 5,948 edges, respectively. The mean degree is 75.41 and the average closeness centrality is 0.1.

4.1   Experiment 1 – Dimensionality of the Reference Layer d

The objective of the first experiment was to evaluate the influence of the dimension of the reference layer (d) on the divergence scores between the original network and its produced twins. As such, the estimation procedure was performed independently for four fixed values of d, i.e. $d \in \{1, 2, 4, 8\}$. Note that it directly affects the Bayesian optimisation component of the proposed method that is used to estimate r.

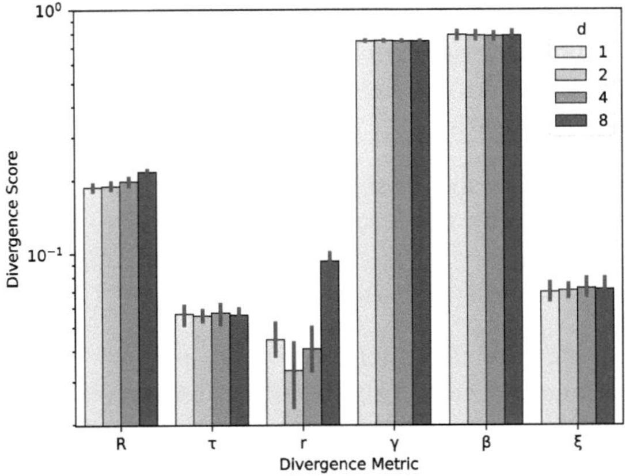

Fig. 3. Divergence scores for configuration retrieval with fixed $d = 2^k$ dimension, $k \in \{0, 1, 2, 3\}$; estimation with tuned r and the optimisation loss $\mathcal{D}_r(\hat{G}, G)$.

Figure 3 presents a comparison of the average values of the corresponding divergence metrics across the tested d. The scores associated with γ, β, ξ, and τ remain virtually unchanged, as expected, since these parameters were fixed during the optimisation of r. Their stability only confirms the reproducibility of the proposed approach. In contrast, a pronounced effect of d is observed for the AMI-based interlayer correlation (r), resulting in non-monotonic divergence values with respect to d, which is consistent with our hypothesis of parameters' interdependence. Nevertheless, the lowest divergence is attained for $d = 2$, which coincides with the design choice recommended by the authors of the mABCD model.

An additional, perhaps less anticipated, effect of using a high d value is the increased divergence in edge correlation (R). The growing misalignment of communities between layers, when comparing the original network to the digital twin, creates unfavourable conditions for efficient link rewiring (phase 6 of the mABCD process). It should also be noted that relatively high values of γ and β can stem from the non necessarily power-law-like distribution of degrees and communities in *Freebase*. A potential improvement for that could be injecting sequences of them directly into mABCD and, by that, to pass over some of its phases.

4.2   Experiment 2 – Configuration Retrieval Methods

The objective of the second experiment was to evaluate various mABCD configuration retrieval methods within the Bayesian framework outlined in Sect. 3. Specifically, we are interested in the following research question. Does increasing the fidelity of the produced twin in one measured dimension trade off for a decrease in other scores (*Open Problem 6*)? To make a step toward this direction, we investigated whether a specific Bayesian tuning configuration exists that outperforms the others across all divergence dimensions.

Based on the considerations, four configuration retrieval runs were executed with the following specifications:

1. decision variable r with loss based on $\mathcal{D}_r(\hat{G}, G)$;
2. decision variables r and τ with loss based on $\mathcal{D}_r(\hat{G}, G)$;
3. decision variables r and τ with loss based on $\mathcal{D}_\tau(\hat{G}, G)$;
4. decision variables r and τ with loss based on $(\mathcal{D}_r(\hat{G}, G) + \mathcal{D}_\tau(\hat{G}, G))/2$.

Based on the outcomes of Experiment 1, we set $d = 2$ for all configuration retrieval specifications. Subsequently, ten digital twins were generated for each set of derived **mABCD** parameters, and the divergence metrics were calculated (Fig. 4).

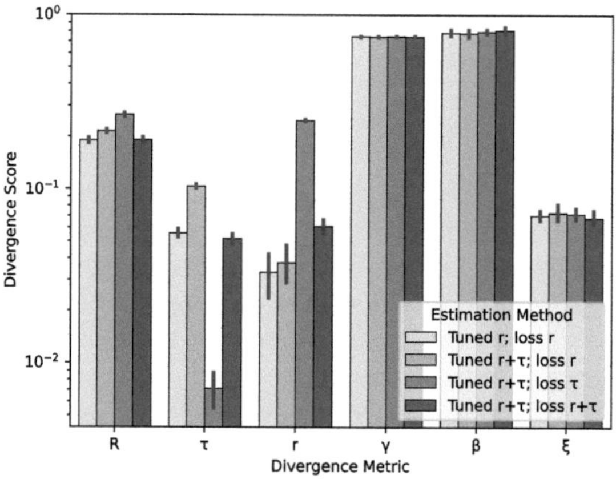

Fig. 4. Divergence scores for four configuration estimation methods; fixed $d = 2$.

The resulting scores for γ, β, and ξ once again validate correct implementation of the configuration retrieval: their minimal variation is attributed to the stochastic nature of **mABCD** and the community extraction algorithm. More significant findings, consistent with expectations, are observed in the divergence metrics $\mathcal{D}_\tau(\hat{G}, G)$ and $\mathcal{D}_r(\hat{G}, G)$. Both scores are minimised when the optimisation technique targets the corresponding loss function (either $\mathcal{D}_\tau(\hat{G}, G)$ or $\mathcal{D}_r(\hat{G}, G)$). However, a trade-off is evident in the form of increased divergence for the alternative metric. This is particularly pronounced for the τ-based loss method, which achieves a divergence score of 0.0071, compared to an average of 0.0701 across other specifications. Conversely, for the $\mathcal{D}_r(\hat{G}, G)$ criterion, the τ-based loss yields 0.2469, indicating suboptimal performance compared to the average of 0.0438 for other losses. Interestingly, expanding the set of decision variables from a single r to include both r and τ (while maintaining $\mathcal{D}_r(\hat{G}, G)$ loss) does not decrease the $\mathcal{D}_r(\hat{G}, G)$ score, but notably exacerbates the $\mathcal{D}_\tau(\hat{G}, G)$.

This phenomenon may be attributed to the curse of dimensionality and insufficient exploration of the parameter space. Variations in the R score are difficult to attribute definitively; however, given their small magnitude, it can be concluded that the specific choice of decision variables and loss functions has a negligible impact on edge correlation fidelity.

To complement the analysis of Experiment 2, we present the loss trajectories for the optimisation model utilising both r and τ as decision variables. Figure 5 illustrates the instantaneous loss calculated at each iteration of the search. It should be noted that we report $\mathcal{D}_\tau(\hat{G}, G)$, $\mathcal{D}_r(\hat{G}, G)$, and $(\mathcal{D}_r(\hat{G}, G) + \mathcal{D}_\tau(\hat{G}, G))/2$ for each loss used as the objective function in the optimisation problem. Consistent reductions in the loss associated with the primary objective are observable, highlighting the efficacy of the Bayesian optimisation. Conversely, the trajectories for the non-target criterion predominantly oscillate around the average, indicating that any reduction in the alternative loss occurs incidentally rather than through directed optimisation.

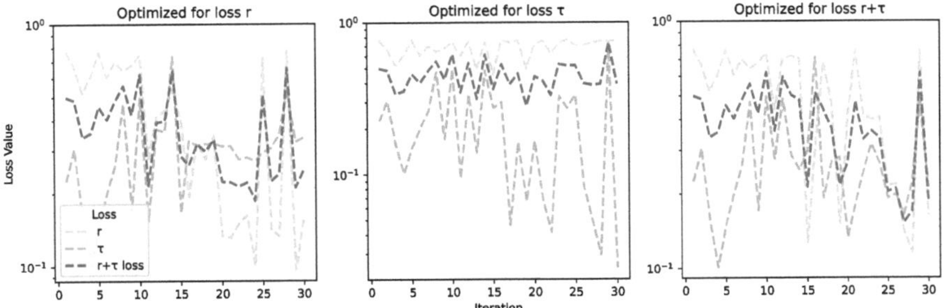

Fig. 5. $\mathcal{D}_\tau(\hat{G}, G)$, $\mathcal{D}_r(\hat{G}, G)$, and $(\mathcal{D}_r(\hat{G}, G) + \mathcal{D}_\tau(\hat{G}, G))/2$ scores as functions of the optimisation step in finding the best matching r or τ.

5 Conclusions

In this work, we discussed a problem of configuration retrieval for multilayer networks with the synthetic network generator—mABCD. Addressing that issue is vital for twinning complex networked systems and opens new possibilities for graph-based data augmentation in machine learning applications.

The main contribution of this study is the proposal of a statistical network inference method from empirical multilayer datasets. We also introduce several divergence measures for assessing the quality of parameter estimation. The method is evaluated on a real-world network representing cooperation within the movie industry. The results provide preliminary evidence supporting the validity of the proposed approach; however, a key finding is that independent parameter estimation techniques encounter intrinsic accuracy limits, as the configuration parameters collectively determine the structural properties of networks generated by mABCD. This observation naturally motivates the development of joint prediction methods as a promising future direction.

Nevertheless, as this work is rather a perspective paper, we left readers with six *Open Problems* that can serve as a roadmap for deeper investigation and which (we believe) will open a fruitful discussion.

OP-1 assessing whether utilising deep embedding methods for joint-parameter estimation is feasible;

OP-2 understanding the influence of the hyper-parameter d and selecting its optimal value;

OP-3 handling networks with degree sequences that do not follow the power-law by injecting them into mABCD;

OP-4 handling networks with community size sequences that do not follow the power-law by injecting them into mABCD;

OP-5 providing a cumulative divergence score, a scalar value which encapsulates all structural discrepancies between the original network and the digital twin;

OP-6 assess if increasing the fidelity of the produced twin with respect to one parameter trades off with a decrease in other scores.

Acknowledgments. This research was partially supported by: (1) EU under the Horizon Europe, grant no. 101086321 OMINO; (2) Polish Ministry of Science and Higher Education, International Projects Co-Funded programme; (3) National Science Centre, Poland, grant no. 2022/45/B/ST6/04145; (4) Polish National Agency for Academic Exchange, Strategic Partnerships programme, grant no. BPI/PST/2024/1/00129/U/00001; (5) Wrocław University of Science and Technology, Academia Profesorum Iuniorum programme. Views and opinions expressed are, however, those of the authors only and do not necessarily reflect those of the founding agencies.

Disclosure of Interests. The authors have no competing interests to declare that are relevant to the content of this article.

Code and Data. The methods presented in this manuscript have been implemented in Python. The source code and installation instructions are available at: https://github.com/network-science-lab/mabcd-for-digital-twins.

References

1. Alstott, J., Bullmore, E., Plenz, D.: powerlaw: a python package for analysis of heavy-tailed distributions. PLOS ONE **9**(1), 1–11 (2014)
2. Barrett, J., DeWolfe, R., Kamiński, B., Prałat, P., Smith, A., Théberge, F.: The artificial benchmark for community detection with outliers and overlapping communities (abcd+o2). In: International Workshop on Modelling and Mining Networks, pp. 125–140. Springer (2025)
3. Barrett, J., Kamiński, B., Prałat, P., Théberge, F.: Self-similarity of communities of the ABCD model. Theoret. Comput. Sci. **1026**, 115012 (2025)
4. Boguna, M., Bonamassa, I., De Domenico, M., Havlin, S., Krioukov, D., Serrano, M.Á.: Network geometry. Nature Rev. Phys. **3**(2), 114–135 (2021)

5. Bródka, P.: A method for group extraction and analysis in multilayer social networks. arXiv preprint arXiv:1612.02377 (2016)
6. Clauset, A., Newman, M.E.J., Moore, C.: Finding community structure in very large networks. Phys. Rev. E **70**(6) (Dec 2004)
7. De Domenico, M., Granell, C., Porter, M.A., Arenas, A.: The physics of spreading processes in multilayer networks. Nat. Phys. **12**(10), 901–911 (2016)
8. Dehghan, A., Prałat, P., Théberge, F.: Network embedding exploration tool (neext). arXiv preprint arXiv:2503.15853 (2025)
9. Easley, D., Kleinberg, J.: Networks, Crowds, and Markets: Reasoning About a Highly Connected World. Cambridge University Press, Cambridge (2010)
10. Henry, A.D., Prałat, P., Zhang, C.Q.: Emergence of segregation in evolving social networks. PNAS **108**(21), 8605–8610 (2011)
11. Kamiński, B., Olczak, T., Pankratz, B., Prałat, P., Théberge, F.: Properties and performance of the ABCDe random graph model with community structure. Big Data Res. **30**, 100348 (2022)
12. Kamiński, B., Pankratz, B., Prałat, P., Théberge, F.: Modularity of the ABCD random graph model with community structure. J. Complex Networks **10**(6), cnac050 (2022)
13. Kamiński, B., Prałat, P., Théberge, F.: Artificial benchmark for community detection (ABCD) - fast random graph model with community structure. Network Sci., 1–26 (2021)
14. Kamiński, B., Prałat, P., Théberge, F.: Artificial benchmark for community detection with outliers (ABCD+o). Appl. Network Sci. **8**(1), 25 (2023)
15. Kamiński, B., Prałat, P., Théberge, F.: Hypergraph artificial benchmark for community detection (h–ABCD). J. Complex Networks **11**(4), cnad028 (2023)
16. Kivelä, M., Arenas, A., Barthelemy, M., Gleeson, J.P., Moreno, Y., Porter, M.A.: Multilayer networks. J. Complex Networks **2**(3), 203–271 (07 2014)
17. Kraiński, Ł, Czuba, M., Bródka, P., Prałat, P., Kamiński, B., Théberge, F.: Multilayer artificial benchmark for community detection (MABCD). Expert Syst. Appl. **307**, 130920 (2026)
18. Lancichinetti, A., Fortunato, S., Radicchi, F.: Benchmark graphs for testing community detection algorithms. Phys. Rev. E **78**(4), 046110 (2008)
19. `skopt.gp_minimize` — scikit-optimize 0.8.1 documentation, October 2021. https://scikit-optimize.github.io
20. Serrano, M.Á., Boguñá, M.: The Shortest Path to Network Geometry: A Practical Guide to Basic Models and Applications. Cambridge University Press (2021)
21. Serrano, M.Á., Krioukov, D., Boguná, M.: Self-similarity of complex networks and hidden metric spaces. Phys. Rev. Lett. **100**(7), 078701 (2008)
22. Tambuscio, M., Ruffo, G., Flammini, A., Menczer, F.: Fact-checking effect on viral hoaxes: a model of misinformation spread in social networks. In: Proceedings of the 24th International Conference on World Wide Web, pp. 977–982 (2015)
23. Wang, X., Liu, N., Han, H., Shi, C.: Self-supervised heterogeneous graph neural network with co-contrastive learning. In: Proceedings of the 27th ACM SIGKDD Conference on Knowledge Discovery & Data Mining, pp. 1726–1736 (2021)
24. Wen, J., Gabrys, B., Musial, K.: DTCNS: a python toolbox for digital twin-oriented complex networked systems. SoftwareX **27**, 101818 (2024)

The Anonymization Problem in Social Networks

R. G. de Jong[1]([✉])[ID], M. P. J. van der Loo[1,2][ID], and F. W. Takes[1][ID]

[1] LIACS, Leiden University, Leiden, The Netherlands
`r.g.de.jong@liacs.leidenuniv.nl`
[2] CBS, Statistics Netherlands, Den Haag, The Netherlands

Abstract. This paper introduces a unified computational framework for the anonymization problem in social networks, where the objective is to maximize node anonymity through graph alterations. We define three variants of the underlying optimization problem: full, partial and budgeted anonymization. In each variant, the objective is to maximize the number of k-anonymous nodes, i.e., nodes for which at least $k-1$ other nodes are equivalent under a particular anonymity measure. We propose four new heuristic network anonymization algorithms and implement these in ANO-NET, a reusable computational framework. Experiments on three common graph models and 19 real-world network datasets yield three empirical findings. First, regarding the method of alteration, experiments on graph models show that random edge deletion is more effective than edge rewiring and addition. Second, we show that the choice of anonymity measure strongly affects both initial network anonymity and the difficulty of anonymization. This highlights the importance of careful measure selection, matching a realistic attacker scenario. Third, comparing the four proposed algorithms and an edge sampling baseline from the literature, we find that an approach which preferentially deletes edges affecting structurally unique nodes, consistently outperforms heuristics based solely on network structure. Overall, our best performing algorithm retains on average 14 times more edges in full anonymization. Moreover, it yields 4.8 times more anonymous nodes than the baseline in the budgeted variant. On top of that, the best performing algorithm achieves a better trade-off between anonymity and data utility. This work provides a foundation for the future development of effective network anonymization algorithms.

Keywords: Complex networks · k-Anonymity · Privacy · Optimization

1   Introduction

In social network analysis, also known as network science, interactions between people are often the object of study for various applications and downstream tasks, including influence maximization [5] and understanding epidemic

© The Author(s), under exclusive license to Springer Nature Switzerland AG 2026
F. C. Graham et al. (Eds.): WAW 2026, LNCS 16630, pp. 18–33, 2026.
https://doi.org/10.1007/978-3-032-27193-8_2

spread [2]. To conduct such research, it is desirable to have large real-world networks of people that are as realistic as possible. However, the identity of individuals in these networks can be revealed based on structural network information, even after removing unique identifiers, i.e., after pseudonymization [12,21]. As a result, sharing or publishing social network data can lead to a breach in the privacy of the people represented. To solve this problem, sometimes also referred to as identity anonymization [18], various works introduced methods to share or publish networks in a privacy aware manner, including clustering [4], differential privacy [10] and k-anonymity [9,17]. Each of these approaches anonymizes the data in a different way. In this work, we focus on k-anonymity methods, which, contrary to approaches mentioned above, allow the creation of an altered anonymized version of the original full network dataset, retaining all individuals.

In k-anonymity, the anonymity of a network as a whole is determined as the fraction of anonymous nodes. A node is k-anonymous if it has, according to a particular *anonymity measure*, the same *signature* as at least $k-1$ other nodes. For example, a node is k-anonymous with respect to the measure of degree if it has the same degree value as $k-1$ other nodes. The choice of measure essentially determines the attacker scenario against which one protects [13].

Much previous work focuses on $k = 2$, measuring network anonymity as uniqueness: the fraction of nodes with a unique signature [12,21]. Given a network and an anonymity measure, the aim of *network anonymization* is then to decrease the uniqueness to a desired level by means of graph alterations, while ensuring that the resulting network data remain suitable for its intended purposes. Consequently, network anonymization involves a trade-off between anonymity and *data utility*. It is a known fact that altering fewer edges leads to higher utility [3]. Utility can also be measured in terms of how well structural network properties are preserved, or as the extent to which performance in network analysis tasks such as community detection is retained after anonymization.

The time complexity of anonymizing a network with as few alterations as possible depends on the chosen anonymity measure. While for degree, anonymization can be done in quadratic time [18], the problem is NP-hard when the exact structure of the 1-neighborhood and node labels are considered [27]. Furthermore, the objective function in the anonymization problem is neither monotonic nor submodular, elegant properties that enabled the development of efficient methods for other common network analysis problems such as influence maximization [5]. Therefore, existing work on the anonymization of networks focuses on (meta)heuristic algorithms, yet always for either a specific anonymity measure [17,27], tailored to a specific type of network [19], or with a focus on random alterations [9,21].

In this paper, we unite these lines of research and introduce a general formulation of the anonymization problem in the form of three meaningful variants, being 1) full, 2) partial and 3) budgeted anonymization. In addition, we consider four commonly used anonymity measures from the literature, providing insights into how the choice of measure affects the anonymization process. For the latter task, we propose four new measure-agnostic heuristic anonymization algorithms.

By means of experiments on both graph models and real-world network data, we independently investigate the effect of A) the chosen alteration operation, B) the anonymity measure, C) the anonymization algorithm and D) the effect of the problem variants on the attained utility. Together with this paper, we release ANO-NET[1], a computational framework for anonymizing networks that enables researchers to implement and evaluate anonymization algorithms across different anonymity measures and different variants of the problem. The framework is accompanied by a representative real-world benchmark dataset.

The remainder of this paper is structured as follows. First, Sect. 2 reviews related work. Section 3 covers preliminaries on networks and anonymization. In Sect. 4, we formally define the anonymization problem and its variants. Section 5 presents our framework for the anonymization process and the proposed heuristic algorithms. We discuss experiments and results in Sect. 6. Finally, Sect. 7 concludes the paper and outlines directions for future work.

2   Related Work

Below, we discuss the most common measures of anonymity and their commonly used algorithms for anonymization (for more details, see for example [17,27]).

Degree. The simplest anonymity measure is the node degree, for which exact algorithms minimizing the number of alterations [17] have been devised.

Neighborhood. Node anonymity measures based on the ego network or d-neighborhood capture more structural information [1,11,21,27]. In node labeled networks, anonymization under this measure is NP-hard [27], making exact approaches infeasible. Consequently, a greedy algorithm that modifies node neighborhoods to resemble those of similar nodes [27], a genetic algorithm [1] and an edge sampling method [21] were proposed.

Graph Invariants. The work of [13] introduced measures based on graph invariants: the number of nodes and edges in the d-neighborhood of a considered node, and the degree distribution within the d-neighborhood.

VRQ. Vertex Refinement Queries (VRQ) [9] consider the multiset of node degrees within distance d of a given node as the measure of anonymity. Anonymization algorithms include random rewiring and algorithms that find similar neighborhoods [25].

In this paper, we contribute to the existing literature by formally introducing and exploring three meaningful variants of the anonymization problem. Instead of proposing algorithms tailored to a specific measure (i.e., a specific attacker scenario), we present a general framework applicable to all variants, and propose several measure-agnostic heuristic anonymization algorithms. Crucially, we empirically examine the impact of different anonymity measures on the anonymization process, and analyze the trade-off between anonymity and utility.

---

[1] https://github.com/RacheldeJong/ANONET.

3   Preliminaries

In this section we formally introduce concepts used throughout this paper, including notions of networks, anonymity and the graph alteration operations.

3.1   Networks

We define a network as an undirected graph $G = (V, E)$ containing a set of nodes V and a set of edges $\{v, w\} \in E$ between nodes $v, w \in V$. The degree of a node is the number of connections it has: $degree(v) = |\{w : \{v, w\} \in E\}|$. The clustering coefficient of a node v equals the number of triangles it is part of, divided by the maximum number of triangles it could form. The distance between two nodes, denoted $dist(v, w)$, is the minimal number of edges that needs to be traversed to move from node v to node w. It follows that $dist(v, v) = 0$ and if no path exists, i.e., if the nodes belong to different components, we let $dist(v, w) = \infty$. The component with the largest number of nodes is called the largest connected component (LCC) or giant component.

The d-neighborhood of a node, $N_d(v) = (V_{N_d(v)}, E_{N_d(v)})$, is the subgraph consisting of all nodes that are at most at distance d of node v, and all edges between them. Two d-neighborhoods are structurally indistinguishable if they are *isomorphic*. We can determine whether two d-neighborhoods are isomorphic by comparing their *canonical labeling* [20], a label assigned by a function $\mathcal{C}(G)$ such that two graphs have the same label value only if they are isomorphic.

Networks often contain communities, groups of nodes more densely connected internally than with the rest of the network. Community detection algorithms [26] assign nodes to communities by favoring intra-community connections and limiting inter-community connections. Some nodes in a network may have more central positions than others. Centrality measures can be used to rank nodes based on their position in the network [2]. The commonly used measure of betweenness centrality assigns a centrality score based on the fraction of shortest paths that pass through the node.

3.2   Anonymity in Networks

Anonymity Definitions. We say that two nodes $v, w \in V$ are equivalent according to anonymity measure M if they have the same signature: $M(v) = M(w)$. We then say that these nodes belong to the same *equivalence class*, denoted $eq_M(v)$ for a node v. A node is k-anonymous if $|eq_M(v)| \geq k$. We let P_M denote the partition of node set V into equivalence classes using measure M. When $k = 2$, the anonymity of a graph can be summarized using *uniqueness*, defined as the fraction of unique nodes in the network with respect to measure M, as shown in Eq. 3. Given this paper's focus on $k = 2$, in this context network anonymity is simply $1 - U(G)$. For later use, Eq. 2 also defines the set of *unique edges* as the edges incident to at least one unique node.

$$V_u = \{v \in V : |eq_M(v)| = 1\} \tag{1}$$

$$E_u = \{\{v, w\} \in E : v \in V_u \vee w \in V_u\} \tag{2}$$

$$U(G) = |V_u|/|V| \tag{3}$$

Alteration Operations. We consider three graph alteration operations:

- **Deletion:** $E' \leftarrow E \setminus \{\{v, w\}\}$, s.t. $\{v, w\} \in E$.
- **Addition:** $E' \leftarrow E \cup \{\{v, w\}\}$, s.t. $\{v, w\} \notin E$.
- **Rewiring:** $E' \leftarrow E \setminus \{\{v, w\}, \{v', w'\}\} \cup \{\{v, w'\}, \{v', w\}\}$
  s.t. $\{v, w\}, \{v', w'\} \in E$ and $\{v, w'\}, \{v', w\} \notin E$.

As preliminary experiments in Sect. 6.2 show that edge deletion is the most effective operation, we focus on edge deletion in the remainder of the text.

Measures for k-Anonymity. Various anonymity measures for k-anonymity have been introduced in the literature, differing both in the structural properties they capture and how far they reach. In this paper, we use the measures listed in Table 1 selecting one representative measure from each category of approaches from previous work covered in Sect. 2. All measures are parameterized by distance d, indicating up to which distance structural information is considered.

Table 1. Anonymity measures (left column), signature for a given node v and distance d (middle column), and node set affected when altering edge $\{v, w\}$ (right column).

Measure	$M(v, d)$	$A_M(\{v, w\})$
DEGREE [17,18]	$degree(v)$	$\{v, w\}$
COUNT [11]	$(\lvert V_{N_d(v)} \rvert, \lvert E_{N_d(v)} \rvert)$	$V_{N_d(v)} \cap V_{N_d(w)}$
d-k-ANONYMITY [21,27]	$\mathcal{C}(N_d(v))$	$V_{N_d(v)} \cap V_{N_d(w)}$
VRQ [9,25]	$\{degree(u) : u \in V,\ dist(u, v) = d\}$	$V_{N_d(v)} \cup V_{N_d(w)}$

To determine its anonymity, each node is assigned a *signature* based on the chosen measure, denoted by $M(v, d)$ in Table 1. When $d \geq 2$, signatures for all $d \geq 1$ up to d should be equal for node equivalence. After an edge is altered, the signature of a subset of the graph's node set changes. Which nodes are affected depends on the *reach* of the anonymity measure, i.e., how far structural information is considered to compute the node signature [13]. The rightmost column in Table 1 lists the set of affected nodes $A_M(\{v, w\})$ after deleting edge $\{v, w\}$, for each measure. This is used later to avoid redundant uniqueness recomputations.

4   The Network Anonymization Problem

We now define the anonymization problem and its three variants in Definition 1.

Definition 1 (Network anonymization problem). *Given a network G, anonymity measure M with range d, a value of k and a set of allowed alteration operations:*

1. ***Full anonymization****: Make all nodes k-anonymous with as few alterations to the network as possible.*
2. ***Partial anonymization****: Ensure that a fraction α (with $0.0 < \alpha < 1.0$) of the nodes is k-anonymous using as few alterations to the network as possible.*
3. ***Budgeted anonymization****: Perform at most B (with $0 < B < |E|$) alterations to the network while maximizing the number of k-anonymous nodes.*

Most existing works focus on full anonymization, aiming to anonymize all nodes in the network. However, some nodes, for example with a high degree as commonly present in real-world data, may require substantially more alterations, motivating the partial variant which generalizes the subset variant proposed in [6]. Since data utility is an important aspect of the problem [17], the budgeted variant limits the number of allowed alterations.

In this paper, we focus on $k = 2$ and $d = 1$, as prior work shows that increasing k (up to 5) has limited effect [12], while $d > 1$ is less realistic and substantially reduces anonymity [13]. Further exploration is left for future work.

5   Approach

In this section we present the general framework for anonymization and introduce the five algorithms (four new, one baseline) used in the remainder of this paper.

5.1   Anonymization Algorithm

The framework outlined in Algorithm 1 can be used for all variants of the anonymization problem introduced in Sect. 4. It takes as input a budget and target anonymity, i.e., the desired number of nodes that is at least k-anonymous. For the full and partial variant, the budget equals $|E|$ and the target anonymity T should be set to $|V|$ and $\alpha * |V|$ respectively, where α equals the fraction of nodes that should be anonymous for the partial variant of the problem. For the budgeted variant, one should give the budget B and a target anonymity of $|V|$. The other inputs required are the original graph G, the measure M and value k, cf. Definition 1. The final two parameters are the chosen anonymization algorithm (see Sect. 5.3), and recompute gap R, which determines after how many edge alterations P_M must be recomputed, further discussed in Sect. 5.2.

The algorithm works as follows. In each iteration of the while loop in lines 3 to 12, B' edges are altered, as selected on line 5 by the chosen anonymization algorithm. The network is altered (line 6) and the affected part of the partition (as defined in the third column of Table 1) is updated in line 7. The remainder of the lines are bookkeeping operations to ensure the right number of edges B', equal to R or the remaining budget if $B < R$, is deleted each iteration and the best result, G_{best}, is updated. The *anonymity()* function returns how many nodes are k-anonymous in the altered graph. When the while loop terminates, either budget B is depleted or target anonymity T is reached. Finally, line 13 returns the graph with the highest anonymity found.

5.2   Recompute Gap

Altering edges changes the network structure, which in turn influences the node signatures, and thus anonymity. After each alteration to the graph, updating the corresponding equivalence classes and data structures is computationally expensive. Therefore, we incorporate a trade-off between accuracy and computational cost by updating the graph and recomputing only after R alterations. With recompute gap $R = 1$ the graph and partition are updated after each alteration, i.e., no recompute gap. For $R = B/100$ the values are updated 100 times.

5.3   Anonymization Algorithms

The heuristic anonymization algorithms described below each assign a probability to edges to be deleted based on specific criteria. We focus on edge deletion as initial experiments in Sect. 6.2 showed that this yielded the best results. This is to be expected, as removing edges reduces the size of node neighborhoods. Smaller neighborhoods admit fewer possible network structures, making these less likely to be unique. The effect of edge deletion is largest on the two incident nodes, which lose one neighbor and all triangles involving that edge. Other affected nodes lose a single edge from their neighborhood. Our structure based heuristic algorithms defined below, leverage these effects. The five proposed heuristic algorithms fall into three categories: random edge sampling; our baseline, two structure based heuristics, and two uniqueness based heuristics.

Random Edge Sampling. The baseline algorithm, ES [21], randomly selects which edges to delete by assigning the same probability to each edge.

$$P_{\text{ES}}(\{v, w\}) = 1/|E| \tag{4}$$

Algorithm 1. ANONYMIZATION

1: **Input: graph** G, **measure** M, **value of** k, **anonymization algorithm** *Anonalg*, **budget** B, **target anonymity** T, **recompute gap** R
2: $G' \leftarrow G$, $G_{best} \leftarrow G$, $P_M \leftarrow compute\_P_M(G, M)$, $P_{Mbest} \leftarrow P_M$
3: **while** $B > 0$ **and** $anonymity(G', P_M, k) < T$ **do**
4:       $B' \leftarrow min(R, B)$
5:       $E_A \leftarrow Anonalg(G', P_M, M, k, B')$ ▷ Select edges to alter (Section 5.3)
6:       $G' \leftarrow update\_graph(G', E_A)$
7:       $P_M \leftarrow update\_P_M(G', P_M, M, E_A)$
8:       $B \leftarrow B - B'$
9:       **if** $anonymity(G', P_M, k) > anonymity(G_{best}, P_{Mbest}, k)$ **then**
10:          $G_{best} \leftarrow G'$, $P_{Mbest} \leftarrow P_M$
11:      **end if**
12: **end while**
13: **Return:** G_{best}

Structure Based. The DEGREE heuristic aims to affect more nodes by assigning higher probabilities to edges connecting two high degree nodes. These edges are likely part of many triangles.

$$P_{\text{DEGREE}}(\{v, w\}) = \frac{min(degree(v), degree(w))}{\sum_{\{v', w'\} \in E} min(degree(v'), degree(w'))} \tag{5}$$

The AFF heuristic computes the exact number of nodes affected by deleting an edge as defined in the rightmost column of Table 1.

$$P_{\text{AFF}}(\{v, w\}) = \frac{|A_M(\{v, w\})|}{\sum_{\{v', w'\} \in E} |A_M(\{v', w'\})|} \tag{6}$$

Uniqueness Based. Since primarily unique nodes need to be altered in order to increase anonymity, this category of algorithms aims to have a larger impact on unique nodes. The first uniqueness-based algorithm, UNIQUE, explicitly targets unique edges, i.e., edges incident to at least one unique node (see Eq. 2). From this set, the desired number of edges B' is chosen at random. If the set of unique edges is insufficiently large, the remaining edges are selected at random from the remaining edge set.

$$P_{\text{UNIQUE}}(\{v, w\}) = \begin{cases} 1/|E_u| & \text{if } \{v, w\} \in E_u, |E_u| > B' \\ 1 & \text{if } \{v, w\} \in E_u, |E_u| \le B' \\ 1/|E \setminus E_u| & \text{if } \{v, w\} \notin E_u, |E_u| < B' \\ 0 & \text{otherwise} \end{cases} \tag{7}$$

The second uniqueness based algorithm, Unique Affected (UA), prioritizes edges whose alteration affects a larger number of unique nodes. To ensure no edge is assigned a selection probability of zero, a factor of $\frac{1}{|E|}$ is added. Similar to the AFF heuristic, this heuristic is more expensive to compute when using the COUNT anonymity measure considered in this paper, as the number of affected nodes need to be computed for each edge.

$$P_{\text{UA}}(\{v, w\}) = \frac{|A_M(\{v, w\}) \cap V_u| + 1/|E|}{\sum_{\{v', w'\} \in E} (|A_M(\{v', w'\}) \cap V_u| + 1/|E|)} \tag{8}$$

5.4   Utility

In this work, we choose to measure data utility based on the change in topological network properties and performance on common network analysis tasks:

- **Clustering coefficient.** The average clustering coefficient over all nodes. This value will likely decrease as triangles are destroyed by deleting edges.
- **Average distance.** The average shortest path length $dist(v, w)$ over all pairs of nodes in the same component. As edges are deleted, paths are destroyed, hence the average shortest path length increases. However, when the network starts to split into components, this value can decrease.

- **Largest connected component (LCC).** The fraction of nodes in the giant component. This is expected to decrease as edges are deleted.
- **Centrality.** Overlap in top 100 most central nodes according to betweenness centrality before and after anonymization. If the top 100 nodes did not change the value equals 1.0, whereas it is 0.0 if there are no common nodes.
- **Community structure.** Normalized Mutual Information (NMI) [15] between the communities found by the Leiden algorithm [26], before and after anonymization. To account for non-determinism of the community detection algorithm, we report on $NMI_{utility} = \max(0.0,\ NMI_{stab} - NMI_{anon})$. Here NMI_{stab} equals the average NMI comparing the community assignments C of the original network, and NMI_{anon} the average NMI between community assignments in the network before (C) and after (C') anonymization; a value of 1.0 implies community structure is preserved, 0.0 that it is completely lost.

6   Results

In this section we first summarize the experimental setup and network datasets. Then, two initial experiments compare the effectiveness of different alteration operations on graph models, and the effect of different anonymity measures on real-world network data. Finally, we compare the proposed anonymization algorithms based on their effectiveness and achieved utility on the three variants.

6.1   Experimental Setup and Data

We implement the algorithms introduced in Sect. 5 in our reusable C++ ANO-NET framework.[2] Experiments on graph models use Python and NetworkX [8], utility analysis is performed with igraph [7]. Experiments are conducted on both graph models and real-world networks. We use three graph models: Erdős-Rnyi (ER), where edges are added at random; Barabási-Albert (BA), accounting for the preferential attachment mechanism that generates a powerlaw degree distribution, and Watts-Strogatz (WS) with rewire probability 0.05, which generates clustered networks [2]. Each model uses graphs with 500 nodes and average degrees 2, 4, 16 and 64. Properties of the real-world network datasets used can be found in Table 2.

In each experiment, edges are sequentially deleted until the desired anonymity level is reached, or the runtime exceeds 30 min. In the latter case, anonymization continues until a budget of $B = 5\%$ of the edges is deleted, corresponding to the budgeted variant. In partial anonymization, we set target anonymity α to 95% of the nodes. The recompute gap is set to delete 1% of the edges in each iteration. To account for nondeterminism in the algorithms and graph models, we report averages and show error bars indicating standard deviations over five runs and five generated graph instances.

For utility analysis, we generate anonymized graphs for the 19 networks that finished within the time limit by deleting the selected edges and compute the utility metrics summarized in Sect. 5.4. For community detection, we run the Leiden algorithm [26] 10 times for each original and anonymized network.

---

[2] https://github.com/RacheldeJong/ANONET.

Table 2. Real-world networks used, listing the number of nodes, edges, average degree, average clustering coefficient, fraction of nodes in the largest connected component (LCC) and average distance. Results include the initial uniqueness using COUNT; values below 0.05 (partially anonymous networks) are shown in italics. The improvement ratio for the three variants is defined as the uniqueness achieved by ES divided by that achieved by UA. Results not included in Sect. 6.4 are indicated by a dash.

	$|V|$	$|E|$	Avg. deg.	Clust. coeff	Frac. LCC	Avg. dist.	$U(G)$	Improvement UA vs. ES		
								1. Full	2. Partial	3. Budgeted
Radoslaw emails [14]	167	3,250	38.92	0.69	1.00	1.97	0.766	8.0	2.3	1.9
Primary school [24]	242	8,317	68.74	0.53	1.00	1.73	0.975	4.5	1.8	1.0
Moreno innov. [14]	241	923	7.66	0.31	0.49	2.47	0.245	6.2	1.8	2.1
Gene fusion [14]	291	279	1.92	0.00	0.08	3.90	*0.024*	4.8	1.0	5.6
Copnet calls [23]	536	621	2.32	0.25	0.65	7.37	*0.024*	5.5	1.0	8.6
Copnet sms [23]	568	697	2.45	0.22	0.80	7.32	*0.026*	2.4	1.0	3.1
Copnet FB [23]	800	6,418	16.05	0.32	1.00	2.98	0.488	7.6	2.1	1.4
FB Reed98 [22]	962	18,812	39.11	0.33	1.00	2.46	0.778	5.3	2.3	1.0
Arenas email [14]	1,133	5,451	9.62	0.25	1.00	3.61	0.230	8.4	1.7	2.4
Euroroads [14]	1,174	1,417	2.41	0.02	0.03	18.37	*0.003*	1.8	1.0	3.7
Air traffic control [14]	1,226	2,408	3.93	0.07	1.00	5.93	*0.042*	5.4	1.0	5.6
Network science [14]	1,461	2,742	3.75	0.88	0.00	5.82	*0.039*	10.0	1.0	44.4
FB Simmons81 [22]	1,518	32,988	43.46	0.33	0.99	2.57	0.785	2.4	2.2	1.4
DNC emails [14]	1,866	4,384	4.70	0.59	0.98	3.37	0.092	15.5	1.5	1.4
Moreno health [14]	2,539	10,455	8.24	0.15	1.00	4.56	0.054	6.7	1.0	3.6
US power grid [14]	3,783	14,124	7.47	0.28	1.00	3.57	*0.008*	5.2	1.0	3.3
Bitcoin alpha [22]	4,941	6,594	2.67	0.11	1.00	18.99	0.113	66.5	1.5	1.9
GRQC collab. [16]	5,241	14,484	5.53	0.69	0.79	6.05	0.054	42.2	1.4	3.1
Pajek Erds [14]	6,927	11,850	3.42	0.40	1.00	3.78	*0.044*	55.7	1.0	4.0
FB GWU54 [22]	12,193	469,528	77.02	0.22	1.00	2.83	0.682	-	-	1.1
Enron email [16]	36,692	183,831	10.02	0.72	0.92	4.03	0.071	-	-	2.0
FB wall 2009 [14]	45,813	183,412	8.01	0.15	0.96	5.60	*0.033*	-	-	4.6
Brightkite [22]	58,228	214,078	7.35	0.27	0.97	4.92	*0.048*	-	-	2.6
Twitter [14]	465,017	833,540	3.59	0.06	1.00	4.59	*0.004*	-	-	1.4

6.2 Alteration Operations and Anonymity

Figure 1 shows results on how repeatedly applying each operation discussed in Sect. 3.2 affects the anonymity of graph models. First note that the number of unique nodes when no edges are altered varies with the average degree. Graphs with higher average degree have a higher initial uniqueness, which corresponds to findings in earlier work [12,21]. Clearly, random deletion is more effective than random addition or rewiring, confirming what we theorized in Sect. 5.3. Interestingly, for the graph models, random addition and rewiring can even result in a higher uniqueness, as shown for the BA model, and the ER and WS models when the average degree is larger than 16. However, for the WS model, with average degree 16, rewiring performs slightly better than deletion when altering a fraction of 0.2 to 0.5 of the edges. This is likely due to the many nearly identical near-cliques in the generated networks.

While these preliminary results can not rule out that edge addition and rewiring can be more effective and perhaps preferred in specific cases, for exam-

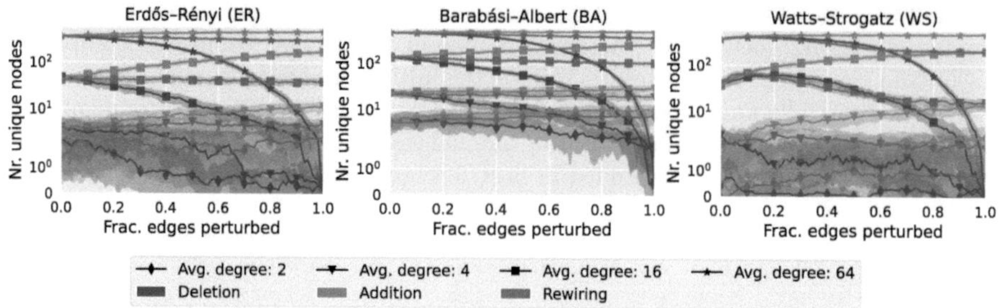

Fig. 1. Random edge deletion (red), addition (green), and rewiring (grey) applied to, using the COUNT measure, anonymize ER (left), BA (middle) and WS (right) graphs with $|V| = 500$ and average degree $\in \{2, 4, 16, 64\}$. (Color figure online)

ple when targeting specific edges, we choose to focus on edge deletion in the remainder of the paper, as it unequivocally performed best on average.

6.3   Anonymization and Measures

Figure 2 shows how uniqueness changes when deleting edges using different measures for anonymity and each algorithm in Sect. 5.3. We depict a selection of five networks representative of the overall observed behavior for the networks listed in Table 2. This is, from top to bottom, left to right in Fig. 2, 1) a plateau with a steep decrease, 2) linear decrease, 3) plateau with linear decrease, 4) sigmoid-like decrease and 5) initial increase before decrease.

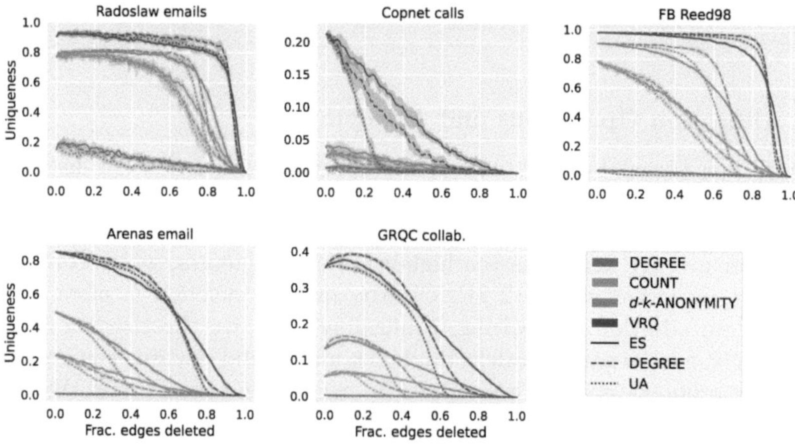

Fig. 2. Uniqueness for four different anonymity measures (color) and three anonymization algorithms (linestyle) on five real-world networks. For each network we show the fraction of deleted edges (horizontal axis) and the attained uniqueness (vertical axis).

Overall, we see that the measure for anonymity used influences both the initial uniqueness and how fast the uniqueness decreases. The simplest measure, DEGREE, has the lowest initial uniqueness and is the easiest to anonymize for. The measure with the furthest reach, VRQ, has the highest initial uniqueness and is the most difficult to anonymize for, corresponding with findings in [13].

Counterintuitively, in the "GRQC collab." network (bottom right), the ES algorithm shows an increase in uniqueness before decreasing. This is likely due to the collaboration network's many cliques. When randomly deleting edges, these cliques are destroyed, which may make certain nodes unique. If edges are targeted more specifically, for example, using the UA heuristic, this initial increase does not occur.

In the remainder of this paper, we focus on the COUNT measure, as this models a realistic attacker scenario, has a substantial initial uniqueness and is easier to anonymize for than, for example, the d-k-ANONYMITY measure.

6.4   Full, Partial and Budgeted Network Anonymization

In this section, we investigate how effective the proposed algorithms from Sect. 5.3 are at solving the three variants of the network anonymization problem.

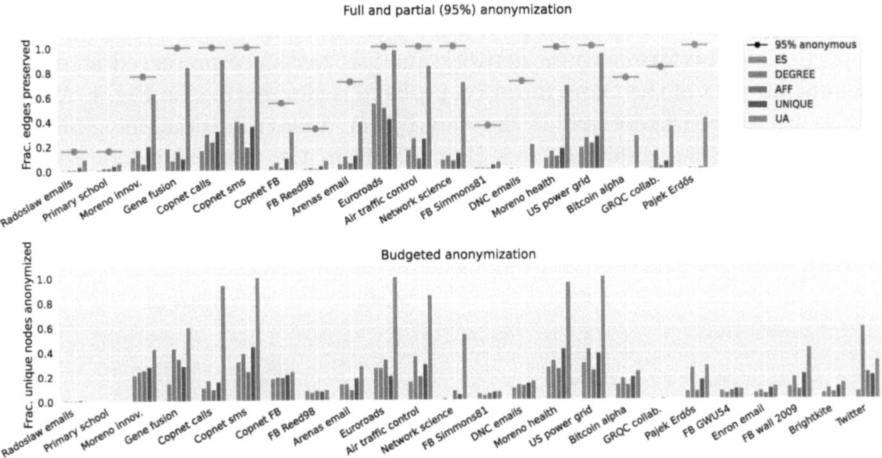

Fig. 3. Top: Results for full and partial anonymization. Bars indicate the fraction of edges preserved in full anonymization, horizontal lines the fraction of edges preserved for partial anonymization. Bottom: Bars indicate the fraction of unique nodes anonymized within the budget of $B = 0.05|E|$. Higher values correspond to better performance.

The top of Fig. 3 shows the fraction of edges preserved $\frac{|E'|}{|E|}$ after deleting edges until the desired level of anonymity is reached. For the partial variant, eight of the networks, including "US Power grid" and "Euroroads" retain 100% of their edges as initially less than 5% of their nodes is unique.

For some networks, such as "Radoslaw emails" and "FB Simmons81" only a small fraction of edges can be preserved. In networks where more edges can be preserved, the uniqueness based algorithm UA outperforms the other algorithms. Dividing the performance of UA by that of ES, we obtain the *improvement ratio* reported in the three rightmost columns of Table 2. On average, for the full and partial variants, 13.9 and 1.8 times more edges can be preserved, suggesting that even simple heuristics provide substantial improvement. The bottom of Fig. 3 shows results for the budgeted variant. To account for differences in initial uniqueness, the vertical axis shows the fraction of unique nodes that is anonymized, $1 - \frac{U(G')}{U(G)}$, when deleting at most 5% of the edges. A higher value indicates greater effectiveness within the given budget.

Overall, the UA algorithm performs well in anonymizing networks, particularly those with lower initial uniqueness, average degree, and higher average distance. Some of the networks, such as "Radoslaw emails" and "GRQC collab.", require more than 5% of edge deletion to achieve substantial improvements. On average, in the budgeted variant, UA anonymizes 4.8 times more nodes than ES.

6.5   Utility

The utility of the anonymized networks in terms of how well various network properties and performance on network analysis tasks are preserved across the different variants of the anonymization problem is summarized in Fig. 4. A property is considered preserved if the value after anonymization, including one standard deviation, differs by less than 5% from the original value. For partial anonymization, we exclude the eight partially anonymous networks (initial uniqueness below 5%), leaving 11 networks.

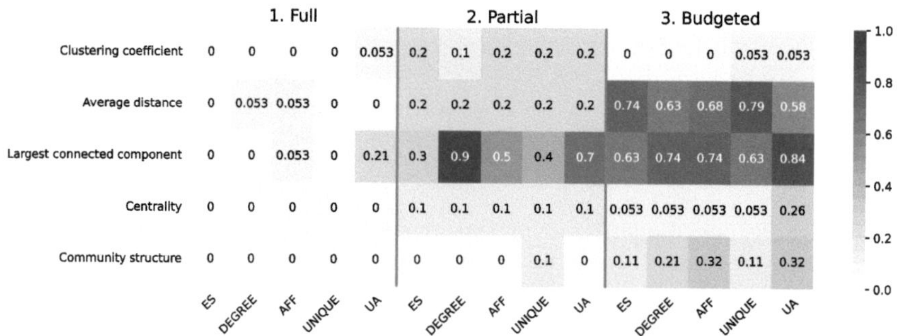

Fig. 4. For each variant of the anonymization problem (columns) and each algorithm (horizontal axis), cells indicate the fraction of networks for which utility (vertical axis) is preserved, i.e., the change ± one standard deviation is less than 5% of the original.

Comparing the three variants, we observe that for full and partial anonymization most properties are generally not preserved. Exceptions are the clustering coefficient and average distance which are preserved for one network, and the largest component size, which the UA algorithm preserves in four networks for the full variant—likely because UA requires fewer edge deletions and targets different edges. In the budgeted variant properties are preserved more often, and differences between algorithms are smaller. The UA algorithm best preserves the majority of properties, except for average shortest path length.

It is also important to note that even random edge deletion (ES) does not always preserve network properties. From the considered properties, average shortest path length and largest component size are preserved most consistently. The latter indicates that the networks often stay connected after anonymization. Community structure and the most central nodes are preserved for a substantial fraction of the networks. However, the average clustering coefficient seems challenging to preserve. These findings suggest that the budgeted variant provides a promising setting for further investigation and comparison of anonymization algorithms when data utility is of importance.

7 Conclusion

In this paper, we introduced the anonymization problem in social networks and its three variants. We introduced ANO-NET, a generic computational algorithmic framework that allows one to anonymize networks under these variants, supporting a range of anonymity measures, the proposed anonymization algorithms and options to explore relevant parameters. Experiments on graph models showed that, among the considered edge alteration operations, random edge deletion is most effective. Moreover, we found that stricter anonymity measures come with higher initial uniqueness and make the anonymization process more challenging. We proposed and evaluated five measure-agnostic anonymization algorithms. Their effectiveness varied across networks; those with lower initial uniqueness and higher average path length were easier to anonymize. The UA algorithm, which targets edges affecting the largest number of unique nodes, consistently showed to be most effective. Compared to the baseline, it anonymized on average 4.8 times more nodes in the budgeted variant and preserved 13.9 and 1.8 times more edges for full and partial anonymization, respectively. Additionally UA better preserved several utility metrics in the budgeted variant, realizing the best overall trade-off between anonymity and data utility.

This work lays the foundation for further studying the anonymization problem and its variants and for extending anonymity measures and algorithms to account for additional network properties such as node labels, layered edges or temporal aspects. Finally, dynamically monitoring utility during the anonymization process to guide edge alterations represents another promising avenue for future research.

References

1. Alavi, A., Gupta, R., Qian, Z.: When the attacker knows a lot: the gaga graph anonymizer. In: Proceedings of the 21st Springer International Conference on Information Security, pp. 211–230 (2019)
2. Barabási, A.L.: Network Science (2016)
3. Borgatti, S.P., Carley, K.M., Krackhardt, D.: On the robustness of centrality measures under conditions of imperfect data. Soc. Networks **28**(2), 124–136 (2006)
4. Campan, A., Truta, T.M.: Data and structural k-anonymity in social networks. In: Privacy, Security, and Trust in KDD, pp. 33–54. Berlin, Heidelberg (2009)
5. Chen, W., Wang, Y., Yang, S.: Efficient influence maximization in social networks. In: Proceedings of the 15th ACM SIGKDD International Conference on Knowledge Discovery and Data Mining, pp. 199–208 (2009)
6. Chester, S., Kapron, B.M., Srivastava, G., Venkatesh, S.: Complexity of social network anonymization. Soc. Network Anal. Mining **3**(2), 151–166 (2013)
7. Csardi, G., Nepusz, T.: The igraph software package for complex network research. Int. J. Complex Syst., 1695 (2006)
8. Hagberg, A.A., Schult, D.A., Swart, P.J.: Exploring network structure, dynamics, and function using networkx. In: Proceedings of the 7th Python in Science Conference, pp. 11–15 (2008)
9. Hay, M., Miklau, G., Jensen, D., Towsley, D., Weis, P.: Resisting structural re-identification in anonymized social networks. In: Proceedings of the VLDB Endowment, vol. 1, pp. 102–114 (2008)
10. Jiang, H., Pei, J., Yu, D., Yu, J., Gong, B., Cheng, X.: Applications of differential privacy in social network analysis: a survey. IEEE Trans. Knowl. Data Eng. **35**(1), 108–127 (2021)
11. de Jong, R.G., van der Loo, M.P.J., Takes, F.W.: Algorithms for efficiently computing structural anonymity in complex networks. ACM J. Exp. Algorithmics **28** (2023)
12. de Jong, R.G., van der Loo, M.P.J., Takes, F.W.: The effect of distant connections on node anonymity in complex networks. Sci. Rep. **14**(1), 1156 (2024)
13. de Jong, R.G., van der Loo, M.P.J., Takes, F.W.: A systematic comparison of measures for k-anonymity in networks. arXiv preprint arXiv:2407.02290 (2024)
14. Kunegis, J.: Konect: the koblenz network collection. In: Proceedings of the 22nd International Conference on World Wide Web, pp. 1343–1350 (2013)
15. Lancichinetti, A., Fortunato, S.: Consensus clustering in complex networks. Sci. Rep. **2**(1), 336 (2012)
16. Leskovec, J., Krevl, A.: Snap datasets: stanford large network dataset collection (2014). http://snap.stanford.edu/data
17. Liu, K., Terzi, E.: Towards identity anonymization on graphs. In: Proceedings of the ACM SIGMOD International Conference on Management of Data, pp. 93–106 (2008)
18. Lu, X., Song, Y., Bressan, S.: Fast identity anonymization on graphs. In: Database and Expert Systems Applications, pp. 281–295 (2012)
19. Malik, A., Adams, B., Hassan, A.: Towards graph-anonymization of software analytics data: empirical study on JIT defect prediction. Empir. Softw. Eng. **29**(4), 76 (2024)
20. McKay, B.D., Piperno, A.: Practical graph isomorphism. II. J. Symb. Comput. **60**, 94–112 (2014)

21. Romanini, D., Lehmann, S., Kivelä, M.: Privacy and uniqueness of neighborhoods in social networks. Sci. Rep. **11**(1), 20104 (2021)
22. Rossi, R.A., Ahmed, N.K.: The network data repository with interactive graph analytics and visualization. In: Proceedings of the 29th AAAI Conference on Artificial Intelligence, pp. 4292–4293 (2015)
23. Sapiezynski, P., Stopczynski, A., Lassen, D.D., Jørgensen, S.L.: The copenhagen networks study interaction data. figshare (2019). https://doi.org/10.6084/m9.figshare.7267433.v1
24. Sociopatters: Datasets (2021). http://www.sociopatterns.org/datasets/
25. Thompson, B., Yao, D.: The union-split algorithm and cluster-based anonymization of social networks. In: Proceedings of the 4th International Symposium on Information, Computer, and Communications Security, pp. 218–227 (2009)
26. Traag, V.A., Waltman, L., Van Eck, N.J.: From Louvain to Leiden: guaranteeing well-connected communities. Sci. Rep. **9**(1), 1–12 (2019)
27. Zhou, B., Pei, J.: The k-anonymity and l-diversity approaches for privacy preservation in social networks against neighborhood attacks. Knowl. Inf. Syst. **28**(1), 47–77 (2011)

Network Analysis and Link Prediction in Competitive Women's Basketball

Anthony Bonato[✉] and Morganna Hinds

Department of Mathematics, Toronto Metropolitan University, Toronto, ON, Canada
`abonato@torontomu.ca`

Abstract. Network structure and its role in prediction are examined in competitive basketball at the team and player levels. Adversarial game outcome networks from the National Collegiate Athletic Association (NCAA) Division I women's basketball from 2021 to 2024 are used to compute the common out-neighbor score and PageRank, which are combined into a low-key leader strength that identifies competitors' influence through structural similarity despite relatively low centrality. This measure is related to changes in NCAA NET rankings by grouping teams into quantiles and comparing average rank changes across seasons for both previous-to-current and current-to-next transitions. Link prediction is then studied using node2vec embeddings across three interaction settings.

For NCAA regular-season game networks, cosine similarity between team embeddings is used as a feature in a logistic regression model to predict March Madness matchups. For the Women's National Basketball Association (WNBA) shot-blocking networks, future directed blocking interactions are predicted via logistic regression on concatenated source-target player embeddings. For WNBA passing networks, region embeddings learned from first-quarter passes are evaluated for their ability to predict subsequent passing connections. Across NCAA and WNBA settings, embedding-based models provide statistically significant evidence that higher-order network structure contains predictive signals for future interactions, while the passing experiment shows weaker predictive performance but yields interpretable similarity patterns consistent with passing feasibility.

1 Introduction

Sports outcome prediction has long been a central topic in both academic research and commercial analytics, driven by applications in performance evaluation, forecasting, and strategic decision-making. Over time, predictive methodologies have evolved from rule-based and statistical approaches to more data-driven techniques. In recent years, machine learning methods, including neural networks, random forests, and decision trees, have become increasingly prominent, attracting significant attention from both researchers and industry professionals [6].

Supported by an NSERC Discovery Grant.

F. C. Graham et al. (Eds.): WAW 2026, LNCS 16630, pp. 34–48, 2026.
https://doi.org/10.1007/978-3-032-27193-8_3

Several recent studies have explored the use of machine learning models for predicting basketball game outcomes. In [9], seven classification machine learning algorithms were evaluated for NBA outcome prediction, with the k-nearest neighbors algorithm achieving the highest accuracy at 60.82%. In [19], the authors construct a network representation in which teams are modeled as nodes connected to their opponents and to their past and future games. A graph convolutional network trained on this structure was shown to improve predictive performance for NBA game outcomes.

Despite the growing use of machine learning techniques, network centrality measures have received comparatively less attention in sports outcome prediction. Notable exceptions include [18], which examined Duke University's men's basketball passing network by representing players as nodes and passes as edges, and using centrality and betweenness to identify influential players. Similarly, [7] proposes using PageRank on an adversarial National Football League network to effectively rank football teams.

The objective of this paper is to investigate the predictive value of network-based measures and embedding techniques in competitive women's basketball. In particular, we examine the effectiveness of centrality measures and the node2vec algorithmic framework in basketball interaction networks. By modeling competitions, passes, and defensive actions as networks, we aim to show that complex network representations can uncover latent structural information that complements traditional predictive features and offers an alternative perspective on sports outcome prediction. In addition to team-level competition networks, we analyze player-level passing and blocking networks to illustrate how these representations capture interaction patterns that are difficult to observe through aggregate statistics alone.

The paper is organized as follows. Section 2 presents the network-theoretic concepts and embedding-based methods that underpin this work. Section 3 describes the data sources, network construction procedures, and experimental design, followed by the presentation of results. Finally, Sect. 4 summarizes the main findings and discusses potential directions for future research.

We consider weighted, directed graphs (or *digraphs*) in the paper. Additional background on graph theory may be found in [16]. For more background on complex networks, see [1].

2   Centralities and Embeddings

To model sports dynamics at both the team and player levels, we represent competitive outcomes and in-game interactions, including passing and blocking, using directed networks that encode relational structure. This network-based perspective enables the application of centrality measures and embedding techniques to quantify influence, similarity, and the underlying structure within the competition.

More formally, an *adversarial network* is a directed graph $G = (V, E)$, where each node represents a competitor, and there is a directed edge $(u, v) \in E$ if u has defeated v in a competition. Edge weights represent the frequency of competitive dominance.

2.1   The CON Score

When predicting competitive outcomes, it is important to identify influential competitors within the network. We adopt a centrality-based measure that quantifies the extent to which a node shares common adversaries with other nodes, which serves as a measure of similarity in competitive behavior and overall influence.

For a given graph G, a node w is a *common out-neighbor* of nodes u and v if $(u, w), (v, w) \in E(G)$. We denote by $\mathrm{CON}(u, v)$ the number of common out-neighbors that u and v share. Furthermore, the *CON score*, or *Common Out-neighbor score* [4], of a node u is defined as the total number of common out-neighbors that u shares with all nodes in G. That is,

$$\mathrm{CON}(u) = \sum_{v \in V(G)} \mathrm{CON}(u, v).$$

For a more in-depth definition of the CON score and its relationship to other centrality measures in network science, see [2].

In basketball competition networks, a node with a high CON score exhibits competitive patterns similar to those of many other nodes. This structural similarity supports informed inference about future performance, as many nodes serve as comparable competitors.

Another commonly used measure of network centrality is *PageRank* (PR). In adversarial networks, PageRank is computed on the reversed-edge network, and nodes with high PageRank values are interpreted as having high centrality. Further details on PageRank and its application to adversarial networks can be found in [1].

2.2   Low-Key Leader Strength

For a deeper perspective on performance prediction, we are interested not only in influential nodes but also in nodes with low centrality. Observe that for any u in a given graph G, we have that $\mathrm{CON}(u)$ is a nonnegative integer and $\mathrm{PR}(u) \in [0, 1]$. To compare the two metrics, we apply *unity-based normalization* to both the PageRank and CON scores. This is done by considering the values $X_1, X_2, \ldots, X_n$ corresponding to the n nodes in the network and defining

$$X_{i,\mathrm{norm}} = \frac{X_i - X_{\min}}{X_{\max} - X_{\min}}.$$

Finally, the *low-key leader strength* [3] of a node, denoted ε_i, is defined as

$$\varepsilon_i = \mathrm{CON}_{i,\mathrm{norm}} - \mathrm{PR}_{i,\mathrm{norm}}.$$

A *low-key leader* is a node with a comparatively high low-key leader strength. In the context of basketball competition networks, a low-key leader is an "underdog:" competitor: one that performs similarly to other influential nodes in the network but is less prominent.

2.3 Node2vec

Node2vec is an algorithmic framework for learning node feature representations that preserve meaningful neighborhood structure, and was first introduced in [8]. Let $G = (V, E)$ be a given graph. We define a mapping $f : V \to \mathbb{R}^d$ that assigns each node a d-dimensional feature representation, or vector embedding. Equivalently, f may be viewed as a $|V| \times d$ matrix. For each node $u \in V$, $N(u) \subseteq V$ denotes a sampled *network neighborhood* generated according to a sampling strategy S. The objective of node2vec is to maximize the likelihood of observing the sampled neighborhood of each node given its embedding. This reduces to a Skip-gram-style formulation, with the context window being the sampled neighborhood, which is then used to optimize f via stochastic gradient ascent.

Node2vec introduces randomized local search procedures to sample multiple neighborhood sets $N(u)$, each of fixed cardinality k, for every node u. The sampling strategy determines the type of similarity encoded in the embeddings. Two important notions of similarity are:

1. *Homophily*: Nodes that belong to the same communities or are highly interconnected should be embedded close together.
2. *Structural Equivalence*: Nodes that share similar structural roles (for example, hubs or bridges) should be embedded close together.

In order to accomplish this, we make use of two sampling strategies:

1. *Breadth-first Sampling* (*BFS*): Samples nodes close (in graph distance) to the current node u, yielding a microscopic view of the local structure. This emphasizes *structural equivalence*, since similar roles can be inferred from local neighborhoods.
2. *Depth-first Sampling* (*DFS*): Samples nodes farther (in graph distance) from the current node u, obtaining a macro-view of the graph. This encourages *homophily* by revealing community structure.

Node2vec interpolates between BFS and DFS through second-order random walks controlled by two parameters, p and q. Consider a random walk in which the most recently traversed edge was from node t to node v. The probability of moving from v to a neighbor x is proportional to

$$\alpha_{p,q}(t, x) \cdot w_{v,x},$$

where $w_{v,x}$ is the edge weight, and

$$\alpha_{p,q}(t, x) = \begin{cases} 1/p & \text{if } d_{tx} = 0, \\ 1 & \text{if } d_{tx} = 1, \\ 1/q & \text{if } d_{tx} = 2. \end{cases}$$

where d_{tx} is the shortest path distance between t and x.

Thus, p acts as a *return parameter*, controlling the likelihood that a random walk revisits the previously visited node. Larger values of p reduce the probability of immediately returning to the previous node, while smaller values of p increase the tendency to backtrack and remain close to the starting node. The parameter q serves as an *in-out parameter*, controlling the likelihood that a random walk explores outward from the previous node rather than remaining within its local neighborhood. Larger values of q bias the walk toward breadth-first exploration, while smaller values of q encourage depth-first exploration.

By tuning p and q, node2vec balances local and global exploration, allowing embeddings to capture both structural equivalence and homophily. Random walks are computationally efficient and enable sample reuse, allowing neighborhoods to be generated for many nodes in a single walk. In this paper, we implemented node2vec with $p = 1$ and $q = 1$, the default values of p and q.

3   Experimental Design and Methods

3.1   NCAA Women's Basketball Game Data

An average of 355 American college women's basketball teams participated in the NCAA Division 1 Women's basketball league from 2021 to 2024. The teams, each belonging to 1 of the 31 conferences, play non-conference games early in the regular season, and conclude the regular season with in-conference games and a tournament.

The *NCAA March Madness tournament* is a single-elimination competition held after the regular season concludes. A total of 68 teams participate and compete for the national title. The 31 winners of the conference tournaments receive an automatic bid to March Madness, thereby securing a reserved spot. The NCAA basketball committee votes on its selections for the other 37 tournament positions. The basketball committee then undertakes an extensive process to create a tournament schedule that aims to achieve relative balance. This entails splitting the teams into four regions based on the committee's rankings, while attempting to separate same-conference teams as much as possible. Figure 1 demonstrates the structure of each region's bracket (that is, the predetermined set of matchups in the single-elimination tournament showing which teams play in each round). The winners of the *Elite Eight* games enter the *Final Four*, where the region 1 winner faces the region 4 winner, and the region 2 winner faces the region 3 winner. Finally, the winners of the Final Four games compete in the *National Championship* game. For further information on the March Madness procedures, refer to [10].

Using a comprehensive dataset from Kaggle [15], we acquired the final scores of every basketball game played within NCAA Division 1 Women's basketball from 2021–2024. Each season averaged 4917 total games. A game data network G was constructed for each season, with nodes representing teams. Scoring data between each pair of teams was totaled throughout their matchups for the entire

Fig. 1. The structure of a March Madness regional bracket.

season, and $(u, v) \in E(G)$ if team u outscored team v across all their head-to-head games in the season, with the total amount outscored represented by edge-weight. Figure 2 shows the network formed from both Regular season and March Madness game data for the 2023–2024 season, consisting of 360 nodes and 3903 edges.

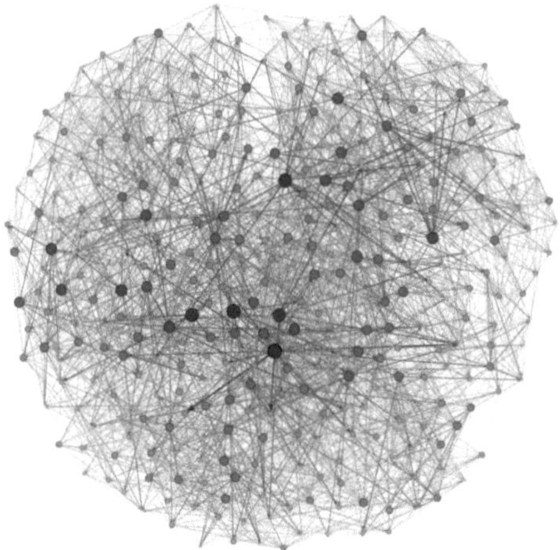

Fig. 2. An NCAA 2023–2024 season network. Nodes with higher comparative out-degree are larger and darker.

3.2 WNBA Passing Data

The *Women's National Basketball Association (WNBA)* is a professional basketball league featuring the world's top female players. Each game consists of four 10-minute quarters, with an additional overtime segment if the score is tied after all quarters. During games, players dribble and pass the basketball to teammates to create open shots. Ball movement is essential to create scoring opportunities for a team. Accordingly, it is imperative that teams agree on pass sequences to optimize scoring. These sequences of passes are called *plays*, and different plays are strategically used throughout the game.

We viewed and manually coded passing data from publicly available WNBA game broadcasts for three WNBA games: Valkyries versus Lynx (09/14/25), Mercury versus Liberty (09/14/15), and Aces versus Fever (09/21/25). Splitting the court into 28 distinct regions, we documented the locations of every pass, both source and target, during these games. The result was an average of 365 passes per game. For each game, we used its corresponding passing data to construct a directed network G, where the nodes are regions of the court, and $(u, v) \in E(G)$ if a pass from region u to region v occurred during the game with $weight(u, v)$ being the number of occurrences. To the best of our knowledge, this is the first time that sports passing data has been represented in this way.

Figure 3 represents the network formed by the 317 total passes made during the Aces vs. Fever game. Note that front-court passes for both teams are tracked on the left-hand side of the network, and back-court passes on the right.

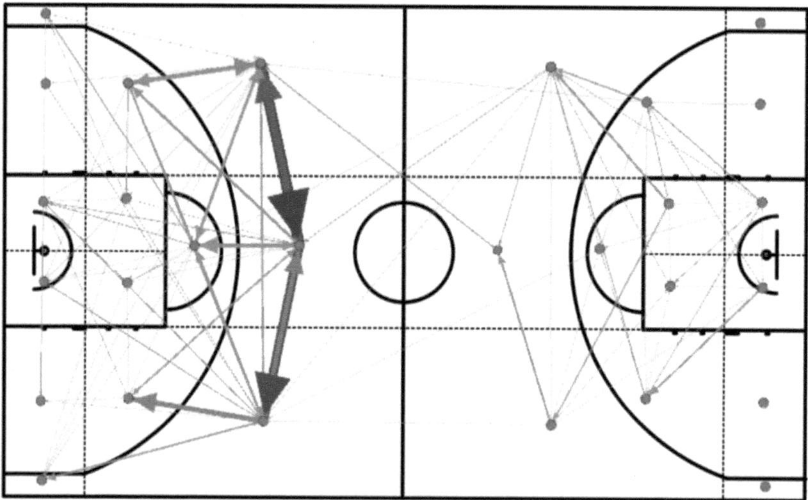

Fig. 3. Aces vs. Fever 09/21/25 passing network. Edges of higher weight are larger and darker in colour.

3.3   WNBA Blocking Data

A crucial defensive tactic in the WNBA is shot-blocking. *Shot-blocking* (or more simply, *blocking*) [17] occurs when a defensive player prevents another player's shot using their arms. This is legal so long as they do not move into the shooter's line of fire. Typically, most blocks are made by players who play the forward position, as they are taller. The source and target of the block depend on many variables, including the court's location and the plays in progress.

Using shot-blocking data from PBP Stats [13], we collected the source and target of every shot block that occurred in the 2023 and 2024 WNBA seasons. There was an average of 1923 blocks per season, with the top-blocking player, A'ja Wilson, having blocked 75 shots in a season. We used this data to form networks G_1 and G_2 for the 2023 and 2024 seasons, respectively. The nodes in each network are players, and $(u, v) \in E(G)$ if player u blocked player v more than vice versa, with $weight(u, v)$ measuring the number of such blocks. Figure 4 shows G_2, the network constructed using data from the 2024 season.

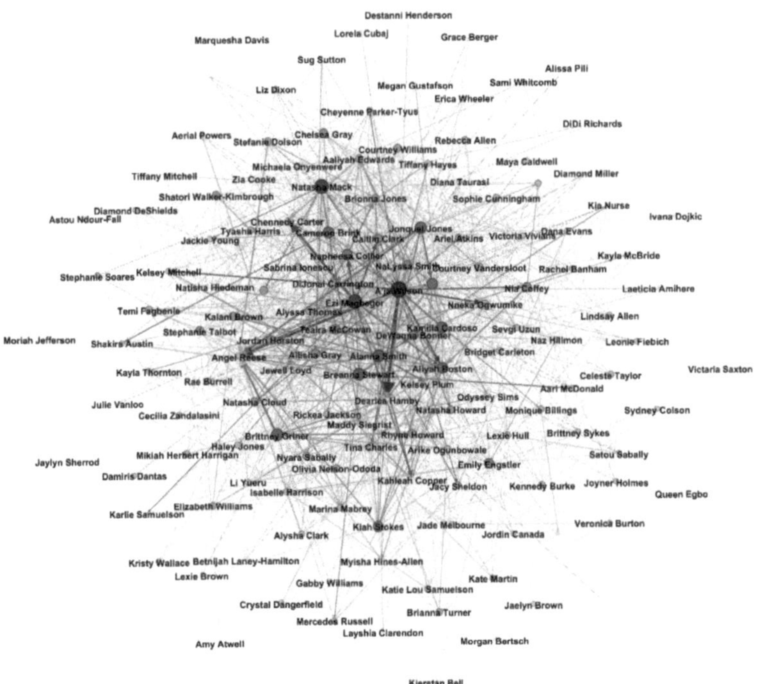

Fig. 4. WNBA 2024 season blocking network. Nodes with higher comparative out-degree are larger and darker.

3.4 Results

Using NCAA women's basketball competition networks from the 20212024 seasons, we computed the CON score, PageRank, and the resulting low-key leader strength for each node. All computations were performed using custom Python scripts developed for this analysis, which are available at github.com/morgannahinds. The resulting low-key leader strengths for NCAA Division I women's basketball teams ranged from -0.361 to 0.949 across the seasons considered.

Low-key leader strengths were then compared with each team's *NCAA Evaluation Tool (NET) ranking*, the NCAA's primary team evaluation metric. Ranking data were scraped from warrennolan.com [14]. For additional details on the methodology used to compute NET rankings, see [11]. We hypothesized that teams with higher relative rankings would show improvements in NET ranking from the previous to the current season, whereas teams with lower relative rankings would experience declines in NET ranking over the same period. Teams were grouped into quantiles, and the average change in ranking from the previous season to the current season was computed for each quantile. The results for the 2024 low-key leader strengths, shown in Fig. 5, are consistent with this hypothesis.

Fig. 5. Average change in NET Ranking from 2023–2024, based on 2024.

We compiled corresponding results for the 2022, 2023, and 2024 seasons. Defining low-key leaders as teams within the interval $[0.4, 1]$, we found that 58 of the 61 quantiles satisfied the stated hypothesis. The same methodology was applied to NCAA men's basketball networks over the same seasons, using game data from [15] and NET rankings from [14]. In this setting, 50 of the

59 quantiles satisfied the hypothesis. The full dataset, including s and average ranking changes, is available at github.com/morgannahinds.

To assess the predictive potential for future rankings, we extended the analysis by evaluating changes in NET ranking between the current and following seasons. We hypothesized that higher would be associated with declines in future NET ranking, whereas lower would be associated with improvements. The results based on the 2022 season are shown in Fig. 6.

Fig. 6. Average change in NET Ranking from 2022–2023, based on 2022.

While the low-key leader analysis focuses on ranking-level prediction, we also investigated whether latent network structure can be leveraged to directly predict future interactions. To this end, we applied the node2vec algorithm to each network to learn low-dimensional node representations for link prediction, embedding each team into a fixed d-dimensional vector space, where the embedding dimension d is a model parameter. In our implementation, we set $d = 128$, the default embedding dimension for node2vec. Network construction and embedding generation were implemented in Python using NetworkX, with node2vec applied via the node2vec Python package [12]. Hypotheses were evaluated by fitting logistic regression models with the statsmodels module, and statistical significance was assessed over 100 independent iterations to account for the stochasticity of the random walks underlying node2vec.

We first applied node2vec to the NCAA game data networks for the 2022, 2023, and 2024 seasons, considering only regular-season games. We hypothesized that teams with similar embeddings would be more likely to face each other in the subsequent March Madness tournament. Embedding similarity was quantified using cosine similarity between the embedding vectors of each team pair. These similarity values served as predictors in a logistic regression model, in which the binary response variable indicated whether the pair of teams played against each other in the subsequent March Madness tournament. The fitted model therefore estimates the probability of a tournament matchup between two teams based on their embedding similarity in the regular-season network. Table 1 reports the average model statistics across all 100 iterations for each season.

Table 1. Logistic regression results using cosine similarity metrics of node2vec embeddings from the regular season game data network to predict March Madness interactions.

Statistic	2022 Value	2023 Value	2024 Value
# team pairs	2152	2135	2142
# embedding features	1	1	1
Pseudo R^2	1.87×10^{-2}	1.2×10^{-2}	3.16×10^{-2}
LLR p-value	2.36×10^{-3}	1.69×10^{-2}	4.16×10^{-5}
Significant embedding dimensions ($p < 0.05$)	1	1	1

Logistic regression models using cosine similarity of node2vec embeddings from regular-season NCAA game networks provide statistically significant evidence that cosine similarity is an effective predictor of March Madness matchups across the 20222024 seasons. Each year, a likelihood ratio test compares a logistic regression model that includes cosine similarity as a predictor with a null model containing only an intercept term. The test indicates that including embedding similarity significantly improves model fit across all years. Although the pseudo-R^2 values are small, this is expected given the sparsity and unpredictability of tournament pairings. Overall, these results suggest that teams that are close together in the regular season are more likely to meet in March Madness.

We next applied the node2vec algorithm to the 2023 WNBA blocking network to determine whether player embeddings could predict future blocking interactions in the 2024 season. Using blocking data from the 2023 WNBA season, we generated 10-dimensional player embeddings. A smaller embedding dimension was used to avoid overfitting, given the relatively small size of the WNBA network. We then tested the predictive utility of these player embeddings by fitting a logistic regression model with the embeddings as predictors of binary blocking outcomes in 2024. Because similar embeddings indicate players with comparable attributes, and blocking interactions are inherently asymmetric, cosine similarity was not appropriate in this setting. Instead, we constructed predictor vectors by concatenating the source and target players' embeddings. Table 2 reports the average summary statistics across all iterations.

Table 2. Logistic regression results using concatenated node2vec embeddings from the 2023 blocking network to predict 2024 blocking interactions.

Statistic	Value
# player pairs	5356
# embedding features	20
Pseudo R^2	0.022
LLR p-value	4.17×10^{-3}
Significant embedding dimensions ($p < 0.05$)	11 of 20

The logistic regression model using concatenated node2vec embeddings from the 2023 blocking network provides statistically significant evidence that the underlying network structure predicts blocking interactions in the 2024 season. The likelihood ratio test strongly rejects the null model ($p = 0.004$), indicating that the embedding-based predictors jointly improve model fit. Again, the pseudo-R^2 is small (0.022), likely due to the sparsity and stochasticity of blocking events. Several embedding dimensions are individually significant, with both positive and negative coefficients, suggesting that multiple structural factors contribute to future blocking behavior. Overall, these findings support the conclusion that higher-order network structure, as captured by node2vec embeddings, contains meaningful information about the formation of future blocking interactions.

We also applied the node2vec algorithm to the WNBA passing network G to assess whether region-level embeddings can predict future passing behavior. In this setting, embeddings were learned from a subgraph of G containing only first-quarter passing data, and we hypothesized that pairs of regions with high cosine similarity would be more likely to be connected by an edge in subsequent quarters. While this approach did not yield strong predictive performance for future pass formations, the resulting similarity scores provided meaningful qualitative insight into the feasibility of passing. Figure 7 illustrates the cosine similarity results of the Fever (09/21/25) for two representative nodes: node 3, the most prominent passing region in the network, and node 13, the least prominent.

Figure 7a clearly categorizes realistic passes originating from region 3. Regions 1 and 5 represent unrealistic passing options due to distance. Although proximity generally increases pass feasibility, several nearby regions present exceptions. Regions 8, 11, 12, 13, and 14 are classified as *high-traffic areas*, meaning that they typically contain multiple players from both teams. This congestion substantially increases the difficulty of passing from region 3 into these areas. In contrast, passes to regions 2, 4, 6, 7, 9, and 10 are more feasible, as these regions are *low-traffic areas* with fewer defenders present. Among these, regions 2, 4, and 9 are particularly favorable due to their proximity and the predominance of right-handed passing mechanics.

Figure 7b illustrates a more challenging passing scenario originating from region 13. Owing to the confined nature of this region and its location within a

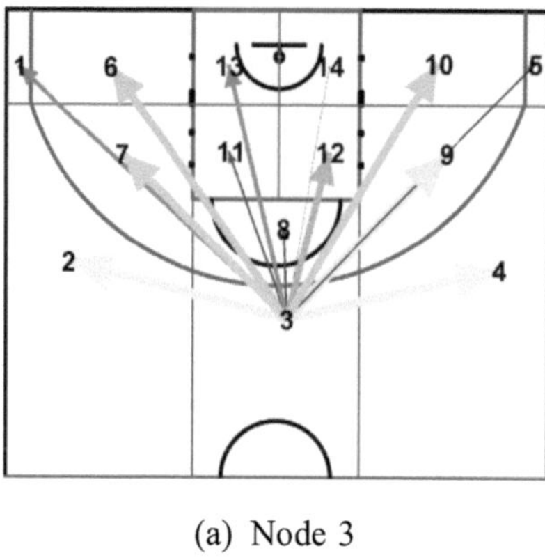

(a) Node 3

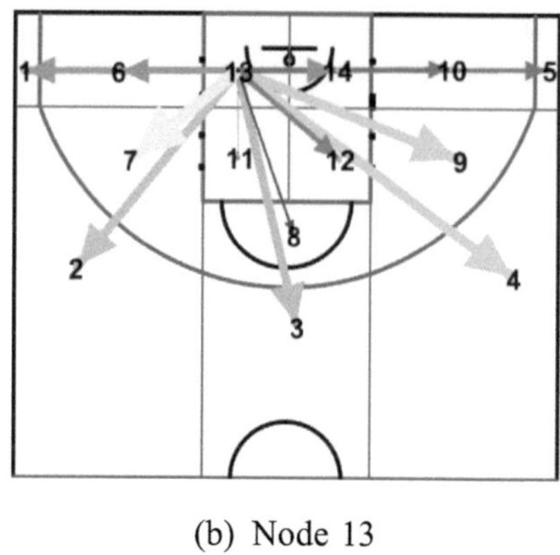

(b) Node 13

Fig. 7. Fever 09/21/25 passing likelihood from nodes 3 and 13 based on cosine similarity of region embeddings. Edges with higher likelihood are larger and lighter in colour. In the Fever passing network, node 3 had the largest out-degree (27) and node 13 had the least out-degree (0).

high-traffic area, passes to nearby high-traffic regions, specifically regions 8, 11, 12, and 14, are likely to be difficult. Passes to regions 1, 5, 6, 10, and 14 are largely infeasible due to the proximity of the out-of-bounds line and the ability of defenders to control the remaining space. In contrast, passes to regions 2, 3, 4, 7, and 9 are more viable options, as these low-traffic regions are sufficiently distant to allow the ball to be lofted over defenders.

4   Discussion and Future Directions

In this work, we demonstrated how network-based representations of competitive and cooperative interactions in basketball can inform performance analysis and support link prediction across collegiate and professional contexts. Using adversarial game outcome networks, we showed that they capture meaningful structural information about changes in NCAA NET rankings. In parallel, we applied node2vec embeddings to multiple basketball networks and found statistically significant evidence that latent network structure predicts future interactions, including postseason NCAA matchups and WNBA shot-blocking behavior, despite the inherent sparsity and stochasticity of these events. While embedding-based prediction was less effective for temporal pass formation in WNBA games, similarity measures still provided interpretable insight into spatial decision-making and passing feasibility. Collectively, these results highlight the value of combining centrality-based measures with embedding-based machine learning approaches to capture higher-order structural signals in sports networks.

From a sports analytics and applied machine learning perspective, these findings indicate that network-derived features, particularly centrality-based measures, can capture underlying competitive information that is not fully reflected in traditional ranking systems. At present, such features remain underutilized in mainstream sports prediction pipelines, which typically emphasize box-score statistics, efficiency metrics, or traditional rating-based models. The low-key leader strength provides a compact, interpretable network feature that can be incorporated alongside standard covariates in team-level predictive models, rather than relying solely on quantile-based analysis as introduced in this paper.

The relationship between low-key leader strength and changes in ranking was also evaluated using college football game outcome networks for the 20222024 seasons. Game data were obtained from the CollegeFootballData API [5], with rankings collected from [14]. Defining low-key leaders as teams within the interval $[0.4, 1]$, we found that 50 of 58 quantiles satisfied the same directional hypothesis observed in NCAA men's and women's basketball. Notably, no violations occurred within the interval $[0.6, 1]$, suggesting that stronger low-key leader signals correspond to more stable predictive behavior.

In contrast, comparable performance was not observed in structurally dissimilar sports such as college baseball and professional tennis. Basketball and football are both physical, team-based sports in which players have specialized roles yet remain free to move throughout the entire playing surface, unlike baseball or tennis, where movement is either highly constrained or individual. This combination of structural freedom and continuous interaction may help explain why centrality measures behave similarly in basketball and football, while proving less effective in college baseball and professional tennis. For the same reasons, these techniques are likely to extend to other structurally similar sports such as soccer, rugby, and hockey.

Model performance may be further improved by integrating contextual variables such as homeaway effects, injuries, roster continuity, or schedule strength. In addition, increased access to fine-grained spatiotemporal data, potentially enabled by computer vision, would enable the application of network embeddings and centrality measures to passing- and play-structure networks at scale. In this setting, machine learning models could jointly leverage offensive and defensive network representations to support matchup-specific prediction, simulation, and strategic analysis.

References

1. Bonato, A.: A Course on the Web Graph, American Mathematical Society, Providence, Rhode Island (2008)
2. Bonato, A., Eikmeier, N., Gleich, D.F., Malik, R.: Centrality in dynamic competition networks, In: Proceedings of the International Conference on Complex Networks and Their Applications (2019)
3. Bonato, A., Kapusin, J., Yuan, J.: Winner does not take all: contrasting centrality in adversarial networks. In: Proceedings of the International Conference on Complex Networks and Their Applications (2024)
4. Bonato, A., Walaa, M.: Analysis and predictability of centrality measures in competition networks, In: Proceedings of the 20th Workshop on Modelling and Mining Networks (WAW 2025) (2025)

5. CollegeFootballData. cfbd-python, GitHub repository. https://github.com/CFBD/cfbd-python
6. Galekwa, R., Tshimula, R., Tshimula, J., Tajeuna, E., Kyamakya, K.: A systematic review of machine learning in sports betting: Techniques, challenges, and future directions, Preprint (2026)
7. Govan, A., Meyer, C., Albright, R.: Generalizing Google's Pagerank to rank national football league teams. In: Proceedings of the SAS Global Forum on Data Mining and Predictive Modeling (2008)
8. Grover, A., Leskovec, J.: node2vec: scalable feature learning for networks. In: Proceedings of the 22nd ACM SIGKDD International Conference on Knowledge Discovery and Data Mining (KDD 2016) (2016)
9. Horvat, T., Havaš, L., Srpak, D.: The impact of selecting a validation method in machine learning on predicting basketball game outcomes. Symmetry **12**, 431 (2020)
10. NCAA. 2024–25 NCAA Division I Women's Basketball Principles and Procedures, NCAA Championships (2024). https://ncaaorg.s3.amazonaws.com/championships/sports/basketball/d1/women/2024-25D1WBB_PrinciplesandProcedures.pdf
11. NCAA. How Do NET Rankings Work in NCAA Tournament Selection?, NCAA News (2025). https://www.ncaa.org/news/2025/3/3/media-center-how-do-net-rankings-work-in-ncaa-tournament-selection.aspx
12. Node2Vec Developers. node2vec: Python implementation of the node2vec algorithm, PyPI package. https://pypi.org/project/node2vec/
13. PBP Stats. PBP Stats: Basketball Play-by-Play Statistics and Tools, online analytics platform. https://www.pbpstats.com/
14. Nolan, W.: WarrenNolan.com: College Basketball Analytics and Rankings, online analytics website. https://www.warrennolan.com/
15. Sonas, J., Mooney, P., Howard, A., Cukierski, W.: March Machine Learning Mania 2025, Kaggle competition dataset (2025). https://kaggle.com/competitions/march-machine-learning-mania-2025
16. West, D.B.: Introduction to Graph Theory, 2nd edn. Prentice Hall (2001)
17. Women's National Basketball Association. WNBA Official Rules 2022, WNBA Rule Book (2022). https://cdn.wnba.com/league/2022/05/2022-WNBA-RULE-BOOK-FINAL.pdf
18. Xu, S.: Duke basketball passing networks: An analysis of player and team dynamics in the context of basketball, Bachelor's thesis, Duke University (2018)
19. Zhao, K., Du, C., Tan, G.: Enhancing basketball game outcome prediction through fused graph convolutional networks and random forest algorithm. Entropy **25**, 765 (2023)

The Iterated Local Model
for Tournaments

Anthony Bonato[1]([⊠]), MacKenzie Carr[1], Ketan Chaudhary[1], Trent Marbach[2],
and Teddy Mishura[1]

[1] Department of Mathematics, Toronto Metropolitan University,
Toronto, ON, Canada
abonato@torontomu.ca
[2] Department of Mathematics and Statistics, Acadia University,
Wolfville, NS, Canada

Abstract. Transitivity is a central, generative principle in social and
other complex networks, capturing the tendency for two nodes with a
common neighbor to form a direct connection. We propose a new model
for highly dense, complex networks based on transitivity, called the Iterated Local Model for Tournaments (ILMT). In ILMT, we iteratively
apply transitivity to form new tournaments by cloning nodes and their
adjacencies, and either preserving or reversing the orientation of existing
arcs between clones. The resulting model generates tournaments with
small diameters and high connectivity as observed in real-world complex
networks. We analyze subtournaments or motifs in the ILMT model and
their universality properties. For many parameter choices, the model generates sequences of quasirandom tournaments. We also study the graph-
theoretic properties of ILMT tournaments, including their cop number,
domination number, and chromatic number. We finish with a set of open
problems and variants of the ILMT model for oriented graphs.

1 Introduction

Complex networks are pervasive in the real world, and key observed properties
include the small-world effect, scale-free degree distributions, and clustering into
dense communities. Such networks are often globally sparse, meaning they have
few edges relative to their number of nodes. See the book [6] for a survey of
complex networks.

In the present work, we consider highly dense networks in which each pair
of nodes is adjacent. *Tournaments* are oriented graphs where each pair of nodes
shares exactly one directed edge (or arc). Tournaments are simplified representations of highly interconnected structures in networks. For example, we may
envision an active sub-thread on the discussion site Reddit, in which users com-
ment on one another's posts. A tournament arises from such social networks by
assigning an arc (u, v) if the user u responds more frequently to the posts of v

A. Bonato—Supported by an NSERC Discovery Grant.

than v does to u. Variants include arcs determined by up- or down-votes. Similarly, densely structured communities also occur on other social media platforms, such as X, Instagram, and TikTok, where subsets of users cluster around a given topic or account. Other examples of tournaments in real-world networks include those in sports, where edges represent one player or team winning over another. Such tournaments on social and other networks grow organically; therefore, it is natural to consider models that simulate their evolution.

Balance theory in undirected graphs cites mechanisms to complete triads (that is, subgraphs consisting of three nodes) in social and other complex networks [18,22]. A central mechanism in balance theory is *transitivity*: if x is a friend of y, and y is a friend of z, then x is a friend of z; see, for example, [27]. Directed networks of ratings or trust scores, along with models for their propagation, were first considered in [19]. *Status theory* for directed networks, first introduced in [23], was motivated by both trust propagation and balance theory. While balance theory focuses on likes and dislikes, status theory posits that an arc indicates that the creator of the link views the recipient as having higher status. For example, on X or other social media platforms, an arc denotes one user following another, and the follower may have higher social status. Evidence for status theory was found in directed networks derived from Epinions, Slashdot, and Wikipedia [23].

Many models simulate the properties of complex networks, such as preferential attachment [2,5]. The *Iterated Local Transitivity* (*ILT*) model introduced in [11] and further studied in [8,12] simulates structural properties in complex networks that emerge from transitivity. Transitivity gives rise to the notion of *cloning*, in which a new node x is adjacent to all the neighbors of an existing node y. Note that in the ILT model, nodes exert local influence within their neighborhoods. The ILT model simulates many properties of social networks. For example, as shown in [11], the model-generated graphs densify over time and exhibit poor spectral expansion. In addition, the ILT model generates graphs with the small-world property, which exhibit low diameter and high clustering coefficient relative to random graphs with the same number of nodes and the same expected average degree. A directed analogue of the ILT model was introduced in [9], and a tournament-based model was first considered in [7]. Other variants of the ILT model include [3,10,13,24].

We introduce a new deterministic model for complex networks based on transitivity that yields tournaments, thereby extending the model considered in [7]. The *Iterated Local Model for Tournaments* (or *ILMT*) model is defined as follows. Fix a *base* tournament G_0 and an infinite binary sequence s named the *generating sequence*. We define a sequence of tournaments G_t (referred to as *time-steps*) as follows.

For each time-step $t \geq 1$, we obtain the tournament G_t from G_{t-1} by the following steps.

1. Add a node x' for each node x of G_{t-1}.
2. Add the arcs $(u, v'), (u', v)$ for each arc (u, v) in G_{t-1}.
3. Add the arc (x', x) for each node x in G_{t-1}.

4. For each arc (u, v) in G_{t-1}, add the arc (u', v') if $s(t) = 1$ or add the arc (v', u') if $s(t) = 0$.

We refer to x' as the *clone* of x in G_t and x as the *parent* of x'. The tournaments G_t are called *ILMT tournaments*. See Fig. 1 for examples of ILMT tournaments.

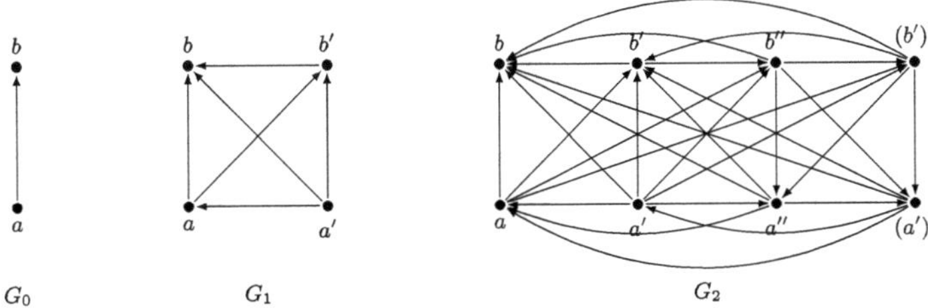

Fig. 1. The first three time-steps of ILMT tournaments, where G_0 is a directed edge, and the sequence s has values $s(1) = 1$ and $s(2) = 0$. In G_2, x'' denotes the clone of x while $(x')'$ denotes the clone of x', for $x = a, b$.

When forming G_t, the *reverse* of G_{t-1} is taken if $s(t) = 0$. Otherwise, the subtournament of G_t induced by the clones is isomorphic to G_{t-1}. We denote G_t more precisely by $\mathrm{ILMT}_{t,s}(G_0)$. For brevity, we often omit explicit references to G_0 and the generating sequence s. We refer to the entries of s that are 0 as its *support*; if there are infinitely many 0-entries, we say that s has *infinite support*.

The paper is organized as follows. In Sect. 2, we prove that ILMT tournaments have low diameter and possess high connectivity. We show that ILMT tournaments are universal in the sense that they embed all finite tournaments. We give precise counts of directed 3-cycles and linear orders of order 3 in ILMT tournaments. Quasirandomness is explored in Sect. 3. If the generating sequence has infinite support, then ILMT tournaments are quasirandom; see Theorem 6. In particular, this gives precise, asymptotic counts of any given finite tournament in ILMT tournaments. In Sect. 4, we consider graph-theoretical properties of ILMT tournaments, including their cop numbers, domination numbers, and chromatic numbers. Our final section includes directions for future research.

Throughout the paper, we consider finite, simple tournaments. A tournament is *strong* if for each pair of nodes x and y, there are directed paths connecting x to y and y to x. The *in-neighbors* and *out-neighbors* of node v are denoted by $N^-(v)$ and $N^+(v)$, respectively. If $x_1, x_2, \ldots, x_n$ are nodes of a tournament G so that (x_i, x_j) is an arc whenever $i < j$, then G is a *linear order*. We denote the linear order on three nodes by T_3 and the directed 3-cycle by D_3. A *sink* is a node with out-degree 0, and a *source* is a node with in-degree 0. For background on tournaments, the reader is directed to [1] and to [28] for background on graphs. For background on social and complex networks, see [6, 15].

2 Diameter, Connectedness, and Motifs

In this section, we show that ILMT tournaments satisfy several properties observed in real-world complex networks, including small diameter and high connectivity. We first show that, under mild restrictions on the base tournament and generating sequence, ILMT tournaments have diameter at most 3.

Theorem 1. *If the tournament G_0 has no sink and s has nonempty support, then for sufficiently large t, the diameter of* $\mathrm{ILMT}_{t,s}(G_0)$ *is at most 3.*

Proof. Let G_0 be a base tournament with no sink and let $G_t = \mathrm{ILMT}_{t,s}(G_0)$, for each $t \geq 1$. In particular, G_0 must have at least three nodes and G_t has no sink for every $t \geq 0$, since a clone at time t has its parent as an out-neighbor.

Suppose $s(t+1) = 0$ for some $t \geq 0$. We show that G_{t+1} must be a strong tournament of diameter at most 3. Pick distinct nodes u and v in G_t such that (u, v) is an arc. Since there are no sinks, there must be a third node w in G_t that is an out-neighbor of v. The nodes u, v', and u' form a D_3 in G_{t+1}. We also have that either the nodes u, v, w' form a D_3 in G_{t+1} or they form a directed 4-cycle along with u', depending on whether w is an in-neighbor or out-neighbor of u, respectively. Altogether, we have that any two nodes in G_{t+1} are in a directed cycle of length at most 4. This is equivalent to the condition that G_{t+1} is strong with diameter at most 3.

Next, we show that if G_t is a strong tournament of diameter at most 3 and $s(t+1) = 1$, then G_{t+1} has diameter at most 3. If $u, v \in V(G_t)$, $\alpha \in \{u, u'\}$, and $\beta \in \{v, v'\}$, then a directed path from α to β can be obtained from a directed path in G_t of length at most 3 from u to v by replacing u with a suitable α and replacing v with suitable β. In this fashion, we have directed paths of length three between all pairs from G_{t+1}. For example, suppose that a directed path from u to v follows from u to x, then to y and terminating at v. A directed path from u' to v then follows from u' to x then y and v. Therefore, G_{t+1} is a strong tournament of diameter at most 3.

Proceeding inductively, we have that if $t_0 \geq 0$ such that $s(t_0 + 1) = 0$, then G_t is a strong tournament of diameter at most 3 for all $t > t_0$. Since the support of s is nonempty, such a t_0 must exist. $\qquad\square$

The *connectivity* $\kappa(G)$ of G is the minimum size of a set of nodes $S \subseteq V(G)$ such that $G \setminus S$ is not strongly connected. A tournament G is called k-*connected* if $\kappa(G) \geq k$. As a consequence of the following theorem, the connectivity of ILMT is unbounded as time-steps become large. The proof is omitted for space considerations.

Theorem 2. *If G_0 is a k-connected tournament, for some $k \geq 1$, and $G_1 =$* $\mathrm{ILMT}_{1,s}(G_0)$ *for some generating sequence s, then G_1 is $2k$-connected.*

A *motif* is an isomorphism type of a specified tournament. Motifs play an important role in the structure of complex networks, both in graphs and in digraphs; see [26]. We provide precise counts of motifs of order 3.

Theorem 3. *Let G_0 be a tournament of order $n_0 \geq 3$, let s be a generating sequence, and for each $t \geq 1$, let $G_t = \mathrm{ILMT}_{t,s}(G_0)$. Let a_t denote the number of subsets of $V(G_t)$ of cardinality 3 that form a D_3 and let b_t denote the number of subsets of $V(G_t)$ of cardinality 3 that form a T_3. We then have that for all $t \geq 1$,*

$$a_t = 2^{s(t)+2} a_{t-1} + (1 - s(t)) \binom{2^{t-1} n_0 + 1}{3},$$

and for all $t \geq 0$,

$$b_t = \binom{2^t n_0}{3} - a_t.$$

Proof. Let $n_t = 2^t n_0$ denote the order of G_t, for each $t \geq 0$. Since each G_t is a tournament, $a_t + b_t$ is equal to the number of subsets of $V(G_t)$ of cardinality 3. Therefore, $b_t = \binom{n_t}{3} - a_t$, for $t \geq 0$. It suffices to show that for $t \geq 1$,

$$a_t = \begin{cases} 8a_{t-1} & \text{if } s(t) = 1, \\ 5a_{t-1} + b_{t-1} + \binom{2^{t-1} n_0}{2} & \text{if } s(t) = 0. \end{cases}$$

Suppose that $s(t) = 1$. No D_3 in G_t can consist of an arc of the form (x', x) for $x \in V(G_{t-1})$. Therefore, every subset of $V(G_t)$ of cardinality 3 that forms a D_3 in G_t is one of the eight sets obtained from the nodes of a D_3 in G_{t-1} by replacing any number of them with their clones. Therefore, $a_t = 8a_{t-1}$.

Next, suppose that $s(t) = 0$. Fix a subset S of $V(G_{t-1})$ of cardinality 3 that forms a D_3. Of the eight sets that one can construct from S by replacing any number of its elements with its clone in G_t, only five form a D_3. If exactly two nodes of S are replaced with their clones, then the subtournament of G_t induced by the resulting set of nodes is a T_3.

If S forms a T_3 in G_{t-1}, then of the eight sets that one can construct from S by replacing any number of its elements with their clones in G_t, we have that seven form a T_3 in G_t. The only one of these eight sets that forms a D_3 in G_t is the one obtained by replacing exactly the source and the sink of T with their clones. Allowing S to range over all cardinality 3 subsets of $V(G_{t-1})$, we have that there are $5a_{t-1} + b_{t-1}$ D_3's in G_t that do not contain an arc from a clone to its parent.

Finally, still under the case $s(t) = 0$, suppose that S is a 3-node subtournament of G_t containing some $x \in V(G_{t-1})$ and its clone x'. We then have that S forms a T_3 if and only if the third node of T is either a neighbor of x in G_{t-1} or the clone of an in-neighbor of x in G_{t-1}. If u is an out-neighbor of x in G_{t-1}, then $\{x, x', u'\}$ forms a D_3. Therefore, the number of cardinality 3 subsets of $V(G_t)$ that form a D_3 and contain both a clone and its parent is equal to the total out-degree of G_{t-1}, which is equal to the size of G_{t-1}. Altogether, we have that

$$a_t = 5a_{t-1} + b_{t-1} + \binom{2^{t-1} n_0}{2},$$

when $s(t) = 0$. Substituting $b_{t-1} = \binom{2^{t-1} n_0}{3} - a_{t-1}$, we obtain the result. $\square$

54   A. Bonato et al.

For a k-node subtournament H of G, its *proportion* is the ratio of the number of copies of H in G, and the number of k-element subsets of $V(G)$. From Theorem 3, we derive the asymptotic proportion of motifs of order 3 in ILMT tournaments. The following theorem does so explicitly for the proportion D_3's, with the proportion of T_3's derived by taking the difference with 1. Note that, as expected, we observe that ILMT typically generates a large proportion of T_3's.

Corollary 1. *Let G_0 be a tournament of order n_0 and let μ be the proportion of D_3's in G_0.*

1. *If $s(t) = 1$ for all $t \geq 1$, then the proportion of D_3's in $\mathrm{ILMT}_{t,s}(G_0)$ approaches $\frac{n_0(n_0-1)(n_0-2)}{n_0^3}\mu$ as $t \to \infty$.*
2. *If s has infinite support, then the proportion of D_3's in $\mathrm{ILMT}_{t,s}(G_0)$ approaches $\frac{1}{4}$ as $t \to \infty$.*

Proof. By Theorem 3 we have that

$$\frac{a_t}{8^t} = 2^{s(t)+2}\frac{1}{8} \cdot \frac{a_{t-1}}{8^{t-1}} + \frac{1-s(t)}{8}\frac{1}{2^{3(t-1)}}\binom{2^{t-1}n_0+1}{3}.$$

Note that $\frac{1}{2^{3(t-1)}}\binom{2^{t-1}n_0+1}{3} \to \frac{n_0^3}{6}$ as $t \to \infty$. Let $\hat{a}_t = \frac{a_t}{8^t}$. We then have that

$$\hat{a}_t = O(2^{-t}) + \begin{cases} \hat{a}_{t-1} & \text{if } s(t) = 1, \\ \frac{\hat{a}_{t-1}}{2} + \frac{n_0^3}{6\cdot 8} & \text{if } s(t) = 0. \end{cases}$$

If s is the constant sequence of 1's, then $\hat{a}_t$ remains constantly a_0, so that the proportion of D_3's $\frac{a_t}{\binom{2^t n_0}{3}}$ approaches

$$\frac{6a_0}{n_0^3} = \frac{n_0(n_0-1)(n_0-2)a_0}{n_0^3\binom{n_0}{3}} = \frac{n_0(n_0-1)(n_0-2)}{n_0^3}\mu$$

as $t \to \infty$. If s has infinite support, then $\hat{a}_t$ approaches the fixed point of the function $x \mapsto \frac{x}{2} + \frac{n_0^3}{48}$, which is $\frac{n_0^3}{24}$. In this case, the proportion of D_3's in G_t approaches $\frac{n_0^3}{24} \cdot \frac{6}{n_0^3} = \frac{1}{4}$. $\square$

If we perform enough 0-steps, then we arrive at the following *universality* property: any motif type can appear eventually in ILMT tournaments after sufficiently many time-steps.

Theorem 4. *Let G_0 be a tournament of order n_0 and let $n \leq n_0$. For a generating sequence s with infinite support, let r be the smallest integer such that*

$$|\{1 \leq t \leq r : s(t) = 0\}| = n.$$

We then have that every tournament on n nodes is isomorphic to a subtournament of $\mathrm{ILMT}_{r,s}(G_0)$.

Proof. Let S_0 be an n-subset of the nodes in G_0, and let G be a tournament with n nodes. Let $\phi_0 : S_0 \to V(G)$ be a bijection. Choose some node u of G. Let $u_0 = \phi_0^{-1}(u)$ and $t_1 = \min\{t : s(t) = 0\}$. Define A_0 to be the set of nodes $v \in S_0$ such that (u_0, v) is an arc in G_0, and $(\phi_0(v), u)$ is an arc in G.

Now, define S_1 to be the n-subset of the nodes of G_{t_1} formed from S_0 by replacing the elements of $A_0 \cup \{u_0\}$ by their clones in G_{t_1}. Let $\phi_1 : S_1 \to V(G)$ be the bijection given by $\phi_1(v) = \phi_0(v)$ for $v \in S_0 \cap S_1$, and $\phi_1(v') = \phi_0(v)$ for $v \in S_0 \setminus S_1$, where v' is the clone of v in G_{t_1}. Now, for every $v \in S_1$, we have that $(\phi_1(u_0'), \phi_1(v)) = (u, \phi_1(v))$ is an arc in G.

We repeat this process of replacing arcs by their reverse, each time choosing a node in G that has not been chosen previously and defining the set S_i and the bijection $\phi_i : S_i \to V(G)$ analogously. After n-many 0-steps, we have that for each arc (x, y) in $\mathrm{ILMT}_{r,s}(G_0)$ with $x, y \in S_n$, $(\phi_n(x), \phi_n(y))$ is an arc in G. $\square$

We have the following corollary.

Corollary 2. *If a generating sequence s has infinite support, then every tournament on n nodes is a subtournament of some ILMT tournament G_t for some $t \geq 0$.*

3 Quasirandomness

Given Corollaries 1 and 2, an interesting question is to estimate the counts of larger motifs. Although this problem appears challenging, we provide an asymptotically exact count for a broad range of ILMT tournaments. To this end, we turn to the theory of quasirandom tournaments, first studied in [17]. Such tournaments are deterministic families that share several asymptotic properties with random tournaments.

Given tournaments G and H, we define the *density* of H in G as the following ratio, where $n(H, G)$ is the number of copies of H present in G, and $\mathrm{Aut}(H)$ is the automorphism group of H:

$$d_G^*(H) = \frac{|\mathrm{Aut}(H)|n(H, G)}{|V(H)|!\binom{|V(G)|}{|V(H)|}}.$$

A sequence of tournaments $(G_n : n \geq 1)$ is said to be a *quasirandom* if for all tournaments H we have that

$$\lim_{n \to \infty} d_{G_n}^*(H) = 2^{-\binom{|V(H)|}{2}}.$$

The uniformly random tournament on n nodes labeled $1, 2, \ldots, n$ has arcs (i, j) with $i < j$ chosen with probability $p = 1/2$. In a quasirandom sequence, the density of any subtournament H in G_n approaches the density of H in the uniformly random tournament. While calculating the density of every subtournament may appear challenging, it may be unnecessary. A tournament G is *quasirandom-forcing* if any sequence of tournaments $(G_n : n \geq 1)$ such that

$$\lim_{n \to \infty} d_{G_n}^*(G) = 2^{-\binom{|V(G)|}{2}},$$

is quasirandom. Quasirandom-forcing tournaments have been completely characterized in the following result.

Theorem 5 ([21]). *If G is a quasirandom-forcing tournament, then G is either a linear order with at least four nodes or a certain non-transitive tournament with five nodes. In particular, a linear order with four nodes T_4 is quasirandom-forcing.*

Hence, to show that a sequence of tournaments is quasirandom, it is sufficient to consider the density of the linear order on four nodes. We take that approach in the following theorem.

Theorem 6. *If s has infinite support, then $(G_t : t \geq 0)$ forms a quasirandom sequence.*

Proof. For simplicity, we consider the sequence consisting of all 0's only. Now, there are four unique tournaments on four nodes: the linear order T_4, the *Winner* tournament, which is the unique non-transitive tournament with a node of out-degree 3, the *Loser* tournament, which is the unique non-transitive tournament with a node of in-degree 3, and the *Mixed* tournament, which is the unique tournament with out-degree sequence $(1, 1, 2, 2)$.

Let G be any tournament, and let G_t be the tournament obtained from t applications of 0-steps to G. Let p_t be the proportion of 4-tuples in G_t that are linear orders, let w_t be the proportion of 4-tuples in G_t that are Winner, let ℓ_t be the proportion of 4-tuples in G_t that are Loser, and let m_t be the proportion of 4-tuples in G_t that are Mixed.

Consider any 4-tuple $X = \{x_1, x_2, x_3, x_4\}$ of nodes in G_t. Each x_i is either a node in G_{t-1} or a clone of a node in G_{t-1}, and so we can create a *corresponding tournament* X^* that is contained entirely in G_{t-1} by replacing each clone node with its parent node in G_{t-1}. The tournament type of X implies the tournament type of X^*; for example, if $X = \{x'_1, x'_2, x'_3, x'_4\}$ forms a Winner tournament in G_t, then $X^* = \{x_1, x_2, x_3, x_4\}$ forms a Loser tournament in G_{t-1}. Thus, for each of the 16 types of 4-node tournaments in G_t, we determine the type of its corresponding tournament in G_{t-1}. Each proportion of the different tournament types in G_t can therefore be expressed as a linear combination of the proportion of tournament types in G_{t-1}, which we state as the following matrix equation:

$$\sigma_t = \begin{bmatrix} p_t \\ w_t \\ \ell_t \\ m_t \end{bmatrix} = \begin{bmatrix} \frac{11}{16} & \frac{1}{16} & \frac{1}{16} & \frac{3}{16} \\ \frac{3}{16} & \frac{9}{16} & \frac{1}{16} & \frac{3}{16} \\ \frac{3}{16} & \frac{1}{16} & \frac{9}{16} & \frac{3}{16} \\ \frac{3}{16} & \frac{1}{16} & \frac{1}{16} & \frac{11}{16} \end{bmatrix} \begin{bmatrix} p_{t-1} \\ w_{t-1} \\ \ell_{t-1} \\ m_{t-1} \end{bmatrix} = T\sigma_{t-1}$$

We note that σ_t only describes the proportion of 4-tournaments X in G_t that do not simultaneously contain a node and its clone. For example, regardless of what type of 4-tournament is formed by the nodes $X = \{x, x', y, z\}$, its corresponding tournament $X^* = \{x, y, z\}$ will not form a 4-tournament. However, the proportion of these tournaments is $\binom{2^t |V(G)|}{3} / \binom{2^t |V(G)|}{4} \sim \frac{4}{2^t |V(G)|}$, which quickly goes to 0 as $t \to \infty$. Therefore, no matter the behavior of these tournaments, they do not meaningfully contribute to the overall proportion of 4-tournament types in G_t. Hence, the above matrix equation fully governs the long-term behavior of tournament types in G_t.

We note that the matrices σ_t form an irreducible aperiodic Markov chain on a finite number of states with transition matrix T. It is well known that each such Markov chain has a limiting distribution π that is also a *stationary distribution* (see, for example, [25, Corollary 7.9]) that is the unique solution to the matrix equation $\pi T = \pi$. Hence, independent of the initial proportions of each of the types of 4-node tournaments in G_0, the proportions σ_t must converge to this distribution π. It is not difficult to show that $\pi^T = [\frac{3}{8} \ \frac{1}{8} \ \frac{1}{8} \ \frac{3}{8}]$, and so as t tends to ∞, the proportion

$$p_t = \frac{n(T_4, G_t)}{\binom{|V(G_t)|}{4}} \to \frac{3}{8}.$$

Hence, we have that

$$d^*_{G_k}(T_4) = \frac{|\mathrm{Aut}(T_4)|}{|V(T_4)|!} p_k = \frac{1}{24} p_t \to \frac{1}{64} = 2^{-\binom{|V(T_4)|}{2}}.$$

The proof follows from Theorem 5. $\qquad\square$

We conclude the section by noting that the ILMT process yields, in a certain sense, continuum-many pairwise non-isomorphic sequences of tournaments. Hence, combined with Theorem 6, the ILMT model yields continuum-many non-isomorphic sequences of quasirandom tournaments.

Theorem 7. *Given a tournament G_0 and two distinct generating sequences s and s', there is some $t \geq 1$ such that $\mathrm{ILMT}_{t,s}(G_0)$ is not isomorphic to $\mathrm{ILMT}_{t,s'}(G_0)$.*

Proof. Let $G_t = \mathrm{ILMT}_{t,s}(G_0)$ and let $G'_t = \mathrm{ILMT}_{t,s'}(G_0)$. If G_1 is not isomorphic to G'_1, then the desired result holds with $t = 1$. It suffices to show that if $t_0 > 0$ such that $G_{t_0} = G'_{t_0}$ and if $s(t_0 + 1) \neq s'(t_0 + 1)$, then G_{t_0+1} is not isomorphic to G'_{t_0+1}.

Assume $t_0 > 0$ such that $G_{t_0} = G'_{t_0}$ and $s(t_0 + 1) = 1 = s'(t_0 + 1) + 1$. Let r be the number of nodes in G_{t_0} with out-degree $2^{t_0-1} n_0$. A node with out-degree δ at time t_0 must have double the out-degree, 2δ, at time $t_0 + 1$. Every clone in G'_{t_0+1} has $2^{t_0} n_0 - 1$ out-neighbors other than its parent. Thus, G'_{t_0+1} has $r + 2^{t_0} n_0$ nodes with out-degree $2^{t_0} n_0$. However, note that G_{t_0+1} has only r nodes with out-degree $2^{t_0} n_0$, since all clones in a 1-step have odd out-degree. Altogether, we have that the tournaments G_{t_0+1} and G'_{t_0+1} are not isomorphic, since they do not have the same out-degree distribution. That is, the desired result holds for $t = t_0 + 1$. $\qquad\square$

4   Graph-Theoretical Properties

We next present results on classical graph-theoretical properties of ILMT tournaments. For a tournament G, we say that $S \subseteq V(G)$ is *in-dominating* in G if every node in $V(G) \setminus S$ has an in-neighbor in S. Out-dominating sets are defined analogously. An in-dominating (respectively, out-dominating) set for G is said to be *minimal* if it does not contain any proper subset that is also in-dominating (respectively, out-dominating) for G. The *in-domination number* of G, denoted $\gamma^-(G)$, is the least cardinality among in-dominating sets for G. The *out-domination number* of G, denoted $\gamma^+(G)$, is defined analogously. In the case that G is a tournament, the condition that $\gamma^-(G) = 1$ is equivalent to the condition that G has a source. Similarly, when G is a tournament, then $\gamma^+(G) = 1$ if and only if G has a sink.

We provide the following lemma.

Lemma 1. *Let G_0 be a tournament, let G_1 be the tournament obtained from G_0 as the result of a 0-step, and for any $S \subseteq V(G_0)$ let $S' = \{x' : x \in S\}$. We then have the following.*

1. $S' = \{x' : x \in S\}$ is an in-dominating set for G_1 if and only if S is both an in- and out-dominating set for G_0.
2. If S with $|S| > 1$ is a minimal in-dominating set for G_0, then S' is an in-dominating set for G_1.

Proof. For space considerations, we only prove (1). If S' is an in-dominating set for G_1, then for every $x \in V(G_0) \setminus S$, there exist some $u, v \in S$ such that (u', x) and (v', x') are arcs in G_1. We have that (u, x) and (x, v) are both arcs in G_0, so S is both in- and out-dominating for G_0.

Conversely, if S is both in- and out-dominating for G_0, then every clone in $V(G_1) \setminus S'$ has an in-neighbor in S' that is an out-neighbor of its parent, and every node in $V(G_0)$ has an in-neighbor in S that is the clone of one of its neighbors in S. Moreover, every element of S' is an in-neighbor of its parent in S. Thus, S' is an in-dominating set for G_1. Item (1) follows. □

Theorem 8. *Let G_0 be a tournament and G_1 the tournament obtained from G_0 as the result of a 0-step. For any $S \subseteq V(G_0)$ with $|S| > 1$, we have that $S' = \{x' : x \in S\}$ is a minimal in-dominating set for G_1 if and only if S is a minimal in-dominating set for G_0.*

Proof. Suppose that S' is a minimal in-dominating set for G_1. By Lemma 1, S is an in-dominating set for G_0. In particular, S contains a minimal in-dominating set for G_0, say S_0. By Lemma 1, the set $S_0' = \{x' : x \in S_0\}$ is an in-dominating set for G_1 that is contained in S'. Since S' is minimal in G_1, we have $S' = S_0'$ and thus, $S = S_0$. Therefore, S is a minimal in-dominating set for G_0.

Conversely, suppose that S is a minimal in-dominating set for G_0. By Lemma 1, S' is an in-dominating set for G_1. Let $S_0 \subseteq S$ such that $S_0' = \{x' : x \in S_0\}$ is a minimal in-dominating set for G_1. By Lemma 1, S_0 is in-dominating for G_0. However, since S is minimal, we have that $S_0 = S$ and so $S' = S_0'$. Therefore, S' is a minimal in-dominating set for G_1. □

We have the following corollary, which shows that the in- and out-domination numbers of ILMT tournaments are either constant across time-steps or eventually constant.

Corollary 3. *Let G_0 be a tournament and $G_t = \mathrm{ILMT}_{t,s}(G_0)$. We have the following:*

1. *$\gamma^+(G_t) = \gamma^+(G_0)$ for all $t \geq 1$.*
2. *If G_0 has no source, then $\gamma^-(G_t) = \gamma^-(G_0)$ for all $t \geq 1$.*
3. *If G_0 has a source and s has infinite support, then $\gamma^-(G_t) = 2$ for sufficiently large t.*

Proof. For $t \geq 1$, if S is an out-dominating set for G_{t-1}, then S is an out-dominating set for G_t. Each clone has its parent as an out-neighbor and, if $v \in V(G_{t-1}) \setminus S$, then v has an out-neighbor in S and so v' does as well. If S is an out-dominating set for G_t, then the set obtained from S by replacing each clone with its parents is an out-dominating set for G_{t-1}. Altogether, we have $\gamma^+(G_{t-1}) = \gamma^+(G_t)$ for all $t \geq 1$. Item (1) follows.

A similar argument shows that $\gamma^-(G_{t-1}) = \gamma^-(G_t)$ when $s(t) = 1$. If G_0 has no source, then G_t also has no source for any $t \geq 1$. By Theorem 8, we have that $\gamma^-(G_{t-1}) = \gamma^-(G_t)$, even when $s(t) = 0$. Item (2) follows.

For item (3), note that if u is a source in G_{t-1} and $s(t) = 1$, then u' is a source in G_t. For some $t_0 \geq 0$, suppose that $s(t_0 + 1) = 0$ and G_{t_0} has a source u. We then have that G_{t_0+1} has no source because every parent has its clone as an in-neighbor, every node other than u and u' has u as its in-neighbor, and u' is a sink among the clones. Since $\{u, u'\}$ is in-dominating for G_{t_0+1}, we then have that $\gamma^-(G_{t_0+1}) = 2$. Applying item (2) to the sequence $s(t + t_0 + 1)$ and base graph G_{t_0+1}, we have that $\gamma^-(G_t) = \gamma^-(G_{t_0+1}) = 2$ for all $t > t_0$. $\square$

We next turn to pursuit-evasion games. *Cops and Robbers* is a game played on a tournament G. There are two players, consisting of a set of cops and a single robber. The game is played over a sequence of discrete time-steps or rounds indexed by non-negative integers, with the cops going first in round 0. The cops and robber occupy nodes. When a player is ready to move in a round, they must move along an arc to a neighboring node. Players can pass or remain on their own node. Any subset of cops may move in a given round. The cops win the game if, after a finite number of rounds, one of them can occupy the same node as the robber. This situation is called a *capture*. The robber wins if they can evade capture indefinitely.

The minimum number of cops required to win is a well-defined positive integer, called the *cop number* of G, written $c(G)$. For additional background on the cop number of a graph, see the book [14].

The cop number in ILMT tournaments does not change with a 1-step, as the following theorem shows (with proof omitted).

Theorem 9. *If G_0 is a tournament and G_1 results from a 1-step, then $c(G_0) = c(G_1)$.*

As the domination number upper bounds the cop number, Corollary 3 gives an upper bound of 2 to the cop number of ILMT tournaments whose generating sequence has infinite support. We show that, even after a single 0-step, the cop number is at most 3.

Theorem 10. *If G_0 is a tournament and G_1 results from a 0-step, then we have that $c(G_1) \leq 3$.*

Proof. Let u and u' be a parent node and its clone in G_1. Suppose that we begin by placing the two cops on u' and the third cop on u. If the robber begins on a clone node v', then either there is an arc (u, v) in G_0, meaning there is an arc (u, v') that the cop on u can use to capture the robber, or there is an arc (v, u) in G_0, meaning there is an arc (u', v') that the cop on u' can use to capture the robber.

We may then assume that the robber starts at a parent node, say v. We then have that either there is an arc (u, v) in G_0, meaning the cop on u can move along this arc to capture the robber, or there is an arc (v, u) in G_0, meaning there is an arc (u', v') that one of the two cops on u' can use to move to the node v'.

Since this is the only case in which the robber is not captured immediately, we assume that this last case occurs. Now, if the robber does not move, the cop on v' can move along the arc (v', v) to capture the robber in the next round. If the robber moves along an arc (v, w), then w must be a parent node to avoid capture by one of the robbers remaining on u and u'. However, there is an arc (v', w), and thus, the cop on v' can move along this arc to capture the robber. $\square$

The following elementary lemma is used to show that the cop number after a 0-step is always at least two.

Lemma 2 ([20]). *If G is a tournament with $c(G) = 1$, then G contains a source.*

As shown in the proof of Corollary 3, an ILMT tournament resulting from a 0-step does not contain a source.

Theorem 11. *If G_0 is a tournament of order at least two and G_1 results from a 0-step, then $c(G_1) \geq 2$.*

We finish the section by considering colorings of ILMT tournaments. A *k-coloring* of a tournament G is a function $\varphi : V(G) \to \{1, 2, \ldots, k\}$ such that the subtournament induced by the set $\{v \in V(G) : \varphi(v) = i\}$ is a linear order, for $1 \leq i \leq k$. The *chromatic number*, written $\chi(G)$, of a tournament G is the minimum k such that G admits a k-coloring.

For a 1-step, the chromatic number remains unchanged.

Theorem 12. *If G_0 is a tournament and G_1 results from a 1-step, then $\chi(G_1) = \chi(G_0)$.*

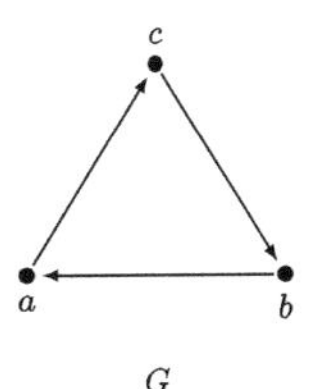
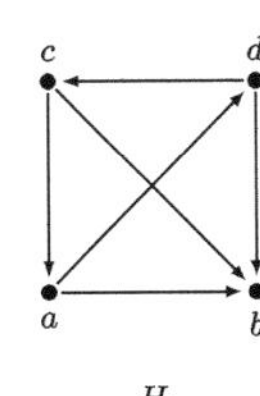
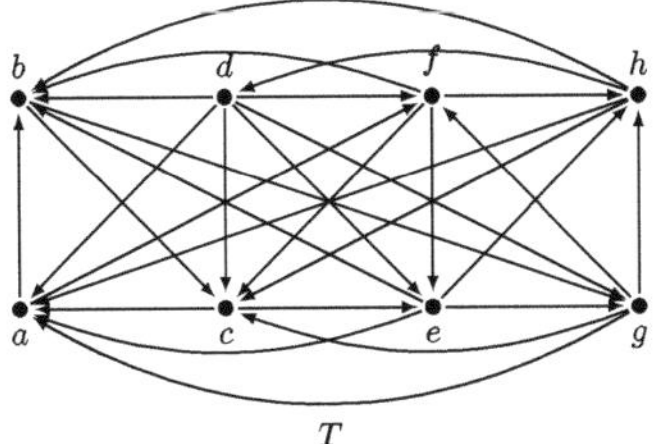

Fig. 2. Tournaments G, H, and T, each with chromatic number 2, for which a 0-step increases the chromatic number by 0, 1, and 2, respectively.

Proof. Let $v_1, v_2, \ldots, v_j$ be a color class in a $\chi(G_0)$-coloring of G_0, labeled so that $v_{\ell+1}, \ldots, v_j$ are out-neighbors of v_ℓ, for $1 \leq \ell < j$. Let $v'_1, v'_2, \ldots, v'_j$ be the clones of $v_1, v_2, \ldots, v_j$, respectively. We then have that $v'_1, v_1, v'_2, v_2, \ldots, v'_j, v_j$ forms a linear order in G_1.

Repeating this for each color class, we obtain a $\chi(G_0)$-coloring of G_1. Hence, $\chi(G_1) \leq \chi(G_0)$. Since G_0 is a subtournament of G_1, we also have that $\chi(G_1) \geq \chi(G_0)$. $\qquad\square$

For a 0-step, we found examples in which the chromatic number remains the same, increases by 1, or increases by 2. See Fig. 2. In general, the chromatic number can at most double as a result of a 0-step. For $t \geq 1$ and $2t$-many 0-steps, we have the following upper bound on the chromatic number of G_{2t}, obtained using a partition of $\mathbb{Z}_2^{2t}$ into sets whose dot products between elements are either always 0 or always 1 (mod 2). By labeling each node of G_{2t} with a binary string of length $2t$ and its parent node in G_0, we can describe the color classes in G_{2t} using color classes in an optimal coloring of the base tournament and the partition of $\mathbb{Z}_2^{2t}$ described above. The full proof is omitted for space considerations.

Theorem 13. *Let G_0 be a tournament and, for $t \geq 0$, let G_{2t} be the tournament obtained by performing $2t$-many 0-steps. We then have that*

$$\chi(G_{2t}) \leq (2^{t+1} - 1)\chi(G_0).$$

We define a sequence S_i of tournaments as follows, first given in [4]. Let S_1 be the one-node tournament. For $i \geq 2$, let $S_i = \Delta(S_{i-1}, S_{i-1}, 1)$ be the tournament formed by taking the disjoint union of two copies of S_{i-1}, call them $S_{i-1}^{(1)}$ and $S_{i-1}^{(2)}$, and a single node v. Add all possible arcs from $S_{i-1}^{(1)}$ to $S_{i-1}^{(2)}$, from $S_{i-1}^{(2)}$ to v, and from v to $S_{i-1}^{(1)}$. Note that S_i contains $2^i - 1$ nodes.

Theorem 14 ([4]). *For $i \geq 1$, $\chi(S_i) \geq i$.*

We have the following corollary, which shows that the chromatic numbers of ILMT tournaments are unbounded.

Corollary 4. *Let $i \geq 1$ and G_0 be a tournament with $n_0 \geq 2^i - 1$ nodes. Let G_t be the tournament obtained by performing t-many 0-steps. We then have that $\log_2(t + 1) \leq \chi(G_t)$.*

Proof. By Theorem 4, at most $t = 2^i - 1$ many 0-steps are required to guarantee that G_t contains the tournament S_i as a subtournament. This subtournament requires $i = \log_2(t + 1)$ colors by Theorem 14, and therefore G_t does as well. $\square$

5   Discussion and Future Work

The ILMT model produces tournaments with small diameters and increasing connectivity. Given a generating sequence with infinite support, the resulting ILMT tournaments contain all motifs as subtournaments. Furthermore, such generating sequences yield continuum-many non-isomorphic quasirandom families of tournaments. We also determined how domination numbers, the cop number, and the chromatic number behave under the 0- and 1-steps in the model.

Several directions remain open. A probabilistic version of ILMT, where each time-step applies a 0- or 1-step with given probability, would be interesting to analyze. The precise growth of the chromatic number under general infinite sequences also remains open. While we have a logarithmic lower bound for the chromatic number of ILMT tournaments, their actual chromatic number is likely much larger. It is unclear how close the upper bound given in Theorem 12 is to being optimal.

While ILMT applies to dense communities in complex networks, a natural extension of ILMT is to sparser oriented graphs. For example, in a 1-step, include the arcs (x', u) for each node x and out-neighbor u of x in G_{t-1}; in a 0-step, include the arcs (x', v) for each node x and in-neighbor v of x in G_{t-1}. See Fig. 3.

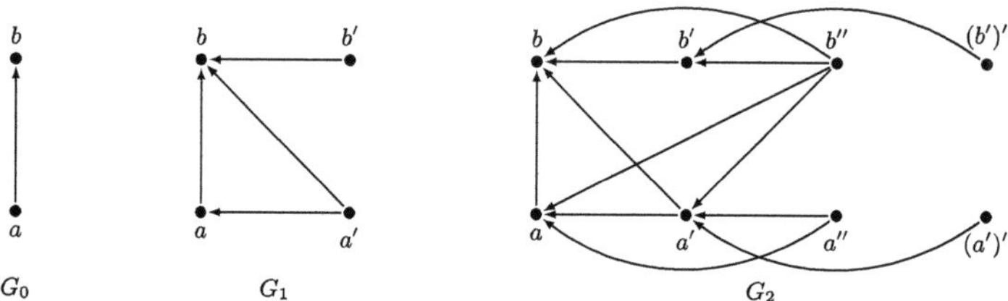

Fig. 3. A sequence of oriented graphs formed by a 1-step and then a 0-step.

Oriented graphs generated by this model will not be tournaments, but will provably generate densifying families. It would be interesting to analyze the complex network and graph-theoretic properties of such iterated oriented models in future work.

References

1. Bang-Jensen, J., Gutin, G.: Digraphs. Springer-Verlag, London Ltd, London (2009)
2. Barabási, A., Albert, R.: Emergence of scaling in random networks. Science **286**, 509–512 (1999)
3. Behague, N., Bonato, A., Huggan, M., Malik, R., Marbach, T.: The iterated local transitivity model for hypergraphs. Discret. Appl. Math. **337**, 106–119 (2023)
4. Berger, E., et al.: Tournaments and coloring. J. Comb. Theory, Ser. B **103**, 1–20 (2013)
5. Bollobás, B., Riordan, O., Spencer, J., Tusnády, G.: The degree sequence of a scale-free random graph process. Random Structures & Algorithms **18**, 279–290 (2001)
6. Bonato, A.: A Course on the Web Graph. American Mathematical Society, Providence, Rhode Island (2008)
7. Bonato, A., Chaudhary, K.: The iterated local transitivity model for tournaments. In: Proceedings of WAW'23 (2023)
8. Bonato, A., Chuangpishit, H., English, S., Kay, B., Meger, E.: The iterated local model for social networks. Discret. Appl. Math. **284**, 556–571 (2020)
9. Bonato, A., Cranston, D.W., Huggan, M.A., Marbach, T., Mutharasan, R.: The iterated local directed transitivity model for social networks. In: Proceedings of WAW'20 (2020)
10. Bonato, A., Cushman, R., Marbach, T., Zhang, Z.: The frustum network model based on clique extension. J. Comb. Optim. **47** (2024)
11. Bonato, A., Hadi, N., Pralat, P., Wang, C.: Models of on-line social networks. Internet Math. **6**, 285–313 (2011)
12. Bonato, A., Infeld, E., Pokhrel, H., Pralat, P.: Common adversaries form alliances: modelling complex networks via anti-transitivity. In: Proceedings of WAW'17 (2017)
13. Bonato, A., Meger, E.: Iterated global models for complex networks. In: International Workshop on Algorithms and Models for the Web-Graph, pp. 135–144 (2020)
14. Bonato, A., Nowakowski, R.J.: The Game of Cops and Robbers on Graphs. American Mathematical Society, Providence, Rhode Island (2011)
15. Bonato, A., Tian, A.: Complex networks and social networks. In: Kranakis, E. (ed.) Social Networks (2011)
16. Chaudhary, K.: The iterated local transitivity model for tournaments. Master's Thesis (2023)
17. Chung, F.R.K., Graham, R.L.: Quasi-random tournaments. J. Graph Theory **15**, 173–198 (1991)
18. Easley, D., Kleinberg, J.: Networks, Crowds, and Markets: Reasoning About a Highly Connected World. Cambridge University Press (2010)
19. Guha, R., Kumar, R., Raghavan, P., Tomkins, A.: Propagation of trust and distrust. In: Proceedings of the 13th International Conference on World Wide Web (2004)
20. Halliday, F.: Cops, robbers and pre-calculus skills. Masters Thesis (2019)
21. Hancock, R., et al.: No additional tournaments are quasirandom-forcing. Eur. J. Comb. **108**, 103632 (2023)
22. Heider, F.: The Psychology of Interpersonal Relations. Wiley (1958)
23. Leskovec, J., Huttenlocher, D., Kleinberg, J.: Signed networks in social media. In: Proceedings of the SIGCHI Conference on Human Factors in Computing Systems (2010)

24. Meger, E., Raz, A.: The iterative independent model. Discret. Appl. Math. **341**, 242–256 (2023)
25. Privault, N.: Long-run behavior of markov chains. In: Understanding Markov Chains: Examples And Applications, pp. 163–188. Springer, Singapore (2018). https://doi.org/10.1007/978-981-4451-51-2_8
26. Milo, R., Shen-Orr, S., Itzkovitz, S., Kashtan, N., Chklovskii, D., Alon, U.: Network motifs: simple building blocks of complex networks. Science **298**, 824–827 (2002)
27. Scott, J.P.: Social Network Analysis: A Handbook. Sage Publications Ltd., London (2000)
28. West, D.B.: Introduction to Graph Theory, 2nd edn. Prentice Hall (2001)

The Needle is a Thread: Finding Planted Paths in Noisy Process Trees

Maya Le[1], Paweł Prałat[2(✉)], Aaron Smith[1], and François Théberge[3]

[1] Department of Mathematics and Statistics, University of Ottawa, Ottawa, Canada
{mle038,asmi28}@uOttawa.ca
[2] Department of Mathematics, Toronto Metropolitan University, Toronto, Canada
pralat@torontomu.ca
[3] Tutte Institute for Mathematics and Computing, Ottawa, Canada
theberge@ieee.org

Abstract. Motivated by applications in cybersecurity such as finding meaningful sequences of malware-related events buried inside large amounts of computer log data, we introduce the "planted path" problem and propose an algorithm to find fuzzy matchings between two trees. This algorithm can be used as a "building block" for more complicated workflows. We demonstrate usefulness of a few of such workflows in mining synthetically generated data as well as real-world ACME cybersecurity datasets.

Keywords: Planted Path Problem · Process Trees · Sequence Recovery

1 Introduction

In many scientific contexts, we observe parts of many large, noisy, labelled directed acyclic graphs and wish to find a small, meaningful path that is common to many of the graphs. In cybersecurity, we might observe the very large "process trees" of compromised computers and attempt to find the sequence of processes used by an attacker. In supply chain management, we might observe the sequence of sites that a large collection of defective or dangerous products and components went to and attempt to find the source of the problem by finding a common path. In biology, we might observe the genealogical trees of cancer cells or viruses and attempt to find the sequence leading to a drug-resistant or virulent version. In software reverse engineering, we might see many call traces and wish to find the common path associated with a pernicious bug.

All of these situations have a common mathematical structure: the main goal is to recover a "planted path" from a large tree. In this paper, we introduce a simple algorithm for extracting paths in Sect. 2.2 and describe several algorithms for incorporating this algorithm into larger machine-learning workflows in Sect. 3. In Sect. 4 we introduce a simple data-generating process and a more formal "planted labelled path" problem that is loosely analogous to the popular "planted clique" problem (see [6]) and related "planted structure" problems (there are many variants; see *e.g.* [2,19]). Throughout the paper, we focus

F. C. Graham et al. (Eds.): WAW 2026, LNCS 16630, pp. 65–81, 2026.
https://doi.org/10.1007/978-3-032-27193-8_5

on developing realistic models and workflows for messy data over developing statistically-optimal solutions for specific clean versions of the "planted path" problem.

While the "planted path" problem occurs in several areas, we were motivated by the problem of finding meaningful sequences of malware-related events buried inside large amounts of computer log data, and we use this as our main illustrative application. Recall that the basic task in cybersecurity triage is to flag a small number of "suspicious" lines of a very large log file for careful inspection by an expert (see *e.g.* the survey [18]). The challenge is that the number of individual log lines that look suspicious under naive heuristics is typically still far too much for human experts to inspect, while the number of "truly bad" events is quite small. However, it is well-known that *certain families* of cybersecurity events (see *e.g.* the sequence from "Reconnaissance" to "Impact" in the MITRE framework[1]) follow our "planted path" structure quite closely.

We acknowledge that the planted path problem is not a perfect abstraction. In real data, the full event might not be strictly contained in a single path and the noisiness of our observations will typically prevent us from observing a full planted path. We argue that the first of these problems is often minor. In Sect. 5 we examine the ACME4 dataset[2] and see that the extracted paths capture a meaningful part of the true event, even though they are not the *full* event. Moreover, a small adjustment to our algorithm would allow us to extract non-paths efficiently in the case that this is necessary. The second problem is more important, but even very imperfect filtering can boost the signal of actually-bad sequences (by aggregating events that are meaningfully related) and filter noise (by removing the vast majority of lines that could not belong to any plausible sequence). It has been widely-recognized that some sort of signal-aggregation is critical for creating statistically powerful detectors [1,14].

This conference paper is an exploratory work, showing that path extraction is both feasible and useful for real cybersecurity data. The journal version will give further details on a more realistic machine learning workflow, comparisons to other algorithms on real datasets, and some basic theory.

2   Fuzzy Matching Algorithm

We start by introducing the notation that will be used in this section.

1. Two directed trees $\mathcal{G}$ and $\mathcal{H}$ on sets of nodes $[n] \equiv \{0, 1, \ldots, n\}$ and, respectively, $[m]$. Nodes 0 are the roots of the corresponding trees. We use $anc_{\mathcal{G}}$ to denote the function mapping a non-root node to its unique ancestor (immediate parent), that is, for any $v \in [n] \setminus \{0\}$, $anc_{\mathcal{G}}(v) \in [n]$. Similarly, $desc_{\mathcal{G}}$ denotes the function mapping a node to its set of descendants, that is, for any $v \in [n]$, $desc_{\mathcal{G}}(v) \subseteq [n]$. In particular, $desc_{\mathcal{G}}(v) = \emptyset$ if and only if v is a leaf. Define $anc_{\mathcal{H}}, desc_{\mathcal{H}}$ analogously for $\mathcal{H}$.

---

[1] https://attack.mitre.org/.
[2] https://gdo168.llnl.gov.

2. A set of features $\mathcal{S}$ and two label functions $\phi_{\mathcal{G}} : [n] \mapsto \mathcal{S}$ and $\phi_{\mathcal{H}} : [m] \mapsto \mathcal{S}$.

3. A weight function $w : \mathcal{S}^2 \mapsto [0, \infty)$.

For clarity of exposition, we assume that each node is characterized by a single feature from the set $\mathcal{S}$. While most real-world datasets store many features for each node, this assumption simplifies the notation but there is no loss of generality. (If *e.g.* a data structure has many features $\mathcal{S}_1, \ldots, \mathcal{S}_k$, we can simply write $\mathcal{S} = \mathcal{S}_1 \times \ldots \times \mathcal{S}_k$.) Within this framework, the weight function w serves to quantify the similarity between any two features.

2.1   Definitions

We introduce the notation used to describe partial matches of *single paths* in each tree. We use the partial ordering of nodes in a tree, formally described as follows:

Definition 1 (Trees and Orderings). *For a directed tree $\mathcal{T}$, define a partial ordering $\leq_{\mathcal{T}}$ on the nodes of $\mathcal{T}$ by writing $u <_{\mathcal{T}} v$ if there is a directed path from u to v.*

Note that there is a unique minimal element, the root of $\mathcal{T}$, but there could be many maximal elements corresponding to leaves of $\mathcal{T}$. We also note that it is always possible to order the set of nodes $[n]$ of $\mathcal{T}$ so that $u <_{\mathcal{T}} v$ implies $u < v$. There are usually many such orderings that can be easily found by, for example, running the Breadth-First Search (BFS) or the Depth-First Search (DFS) algorithms and using the discovery time to order nodes. Let us note that the reverse implication does not hold unless $\mathcal{T}$ is in fact a single path.

For the remainder of the paper, we assume that the nodes of considered trees $\mathcal{G}$ and $\mathcal{H}$ are ordered in the above way. Our goal is to find an optimal matching of nodes in our two trees (with respect to the fixed weight function w). Formally, we introduce the following definition:

Definition 2 (Valid (Partial) Matchings). *We say that a sequence of pairs of nodes $(u_1, v_1), \ldots, (u_k, v_k) \in [n] \times [m]$ is a valid matching between trees $\mathcal{G}$ and $\mathcal{H}$ if $u_i <_{\mathcal{G}} u_{i+1}$ and $v_i <_{\mathcal{H}} v_{i+1}$ for all $i \in \{1, 2, \ldots, k-1\}$.*

We say that such a sequence is a valid partial matching up to $(u, v) \in [n] \times [m]$ if it is a valid matching and furthermore $u_k \leq_{\mathcal{G}} u$ and $v_k \leq_{\mathcal{H}} v$. Let us denote by $\mathcal{P}_{u,v}$ the set of valid partial matchings up to (u, v) and by $\mathcal{P} = \bigcup_{u,v} \mathcal{P}_{u,v}$ the set of all valid matchings.

With the above definition at hand, we can then define the *score* of a valid matching $p = (u_1, v_1), \ldots, (u_k, v_k)$ as follows:

$$s(p) = \sum_{i=1}^{k} w(\phi_{\mathcal{G}}(u_i), \phi_{\mathcal{H}}(v_i)). \tag{1}$$

Finally, the *similarity score* between trees $\mathcal{G}$ and $\mathcal{H}$ is simply the largest possible score over all valid matchings:

$$s(\mathcal{G}, \mathcal{H}) = \max_{p \in \mathcal{P}} s(p). \tag{2}$$

2.2 Matching Algorithm

In this subsection, we give a "fuzzy matching" algorithm for computing $s(\mathcal{G}, \mathcal{H})$, the feature-based similarity score between $\mathcal{G}$ and $\mathcal{H}$ introduced above—see Algorithm 1. The algorithm is very similar to the dynamic programming algorithm of [16]. In particular, we will use the bottom-up (tabulation) variant of the problem which avoids recursion. This method solves a problem by starting with its smallest subproblems and building up solutions to progressively larger subproblems until the original problem is solved. To that end, it uses two arrays, $A_{u,v}$ and $C_{u,v}$, to store the results of these subproblems, avoiding the overhead of recursion associated with the top-down (memorization) approach. The time complexity of this algorithm is clearly polynomial: $O(nm)$.

Throughout this subsection, we use the convention that the root node 0 has a unique ancestor -1, which is never matched. This convention is used purely to avoid introducing special corner cases to the algorithm. Indeed, the algorithm is initialized with $A_{-1,-1} = A_{-1,v} = A_{u,-1} = 0$ for all $u \in [n]$, $v \in [m]$, which corresponds to "empty matches", and then all cases are handled the same way.

Algorithm 1. Basic Matching Algorithm

Input: Trees $\mathcal{G}, \mathcal{H}$ on sets of nodes $[n]$ and $[m]$, label functions $\phi_{\mathcal{G}} : [n] \mapsto \mathcal{S}$, $\phi_{\mathcal{H}} : [m] \mapsto \mathcal{S}$, weight function $w : \mathcal{S}^2 \mapsto [0, \infty)$.

1 Initialize $A_{-1,-1} = A_{-1,v} = A_{u,-1} = 0$ for all $u \in [n]$, $v \in [m]$.

2 **for** $u \in [n]$: **do**

3   **for** $v \in [m]$: **do**

4   Set

$$A_{u,v} = \max \left\{ \begin{array}{l} A_{anc_{\mathcal{G}}(u),v} \\ A_{u,anc_{\mathcal{H}}(v)} \\ w(\phi_{\mathcal{G}}(u), \phi_{\mathcal{H}}(v)) + A_{anc_{\mathcal{G}}(u),anc_{\mathcal{H}}(v)} \end{array} \right\}$$

and $C_{u,v} \in \{1, 2, 3\}$ according to which of these three options was selected (breaking ties in favour of the largest option).

5   **end for**

6 **end for**

7 Initialize $\gamma_{\mathcal{G}}, \gamma_{\mathcal{H}}$ to be empty lists and $\ell = \text{argmax}_{(u,v)} A_{u,v}$.

8 **while** $\ell_1 \neq -1$ and $\ell_2 \neq -1$ **do:**

9   **if** $C_\ell = 3$ **then:**

10   Prepend ℓ_1 to $\gamma_{\mathcal{G}}$ and ℓ_2 to $\gamma_{\mathcal{H}}$.

11   **end if**

12   Set

$$\ell = \left\{ \begin{array}{ll} (anc_{\mathcal{G}}(\ell_1), \ell_2), & C_\ell = 1 \\ (\ell_1, anc_{\mathcal{H}}(\ell_2)), & C_\ell = 2 \\ (anc_{\mathcal{G}}(\ell_1), anc_{\mathcal{H}}(\ell_2)), & C_\ell = 3. \end{array} \right.$$

13 **end while**

14 Let L be the length of $\gamma_{\mathcal{G}}$. Return the sequence $a_{1:L} = (\gamma_{\mathcal{G}}(1), \gamma_{\mathcal{H}}(1)), \ldots, (\gamma_{\mathcal{G}}(L), \gamma_{\mathcal{H}}(L))$ and the score $A = \max_{u,v}(A_{u,v})$.

Figure 1 illustrates a matching of two labelled trees (sampled according to the procedure in Algorithm 4 from the next section) that were matched with Algorithm 1. The sequences in the "planted" paths are independent noisy samples from the ground-truth sequence 4-4-2-4-0-2-2-2-3-1 and are placed on the right-hand sides of the trees. The optimal match in the observed data is the sequence 4-0-2-4-2-2-2-4-3-1, and the matched nodes are exactly along the true planted path. Unmatched nodes are coloured grey; matched nodes in the two trees are coloured with same colour. Note that the "alphabet" of symbols at each node is very small and the planted sequence has noise, so there are both some coincidental "false positive" matches and some accidental "false negative" matches along the ground-truth planted path. Nonetheless, the highest-scoring match is a subset of the path with the planted sequence, which is on the right-hand side of the trees as displayed.

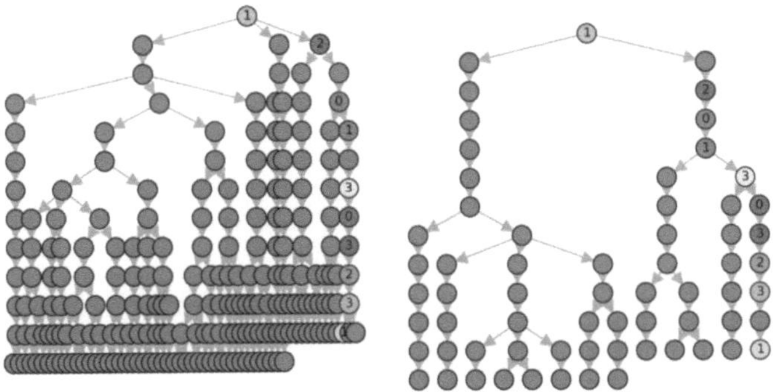

Fig. 1. Example of a matching of two trees.

Remark 1 (When do we expect to find the "ground truth" path?). It might be surprising that we perfectly recover the planted path in Fig. 1, even though the actual matched sequence is not very close to the ground-truth planted sequence. To aid intuition, we give a rough heuristic argument for when we expect the highest-scoring match to be (nearly) a subset of the planted path.

Assume that we have a depth-d m-ary tree in which a given length-d sequence is planted; other node labels are chosen i.i.d. uniformly from an alphabet of size k. Also assume the following simple noise model: labels in the planted path are re-sampled uniformly from the alphabet with some "error" probability $0 < p < 1$.

Let us now consider the optimal match between two randomly labelled trees. Each node along the planted path matches with probability $q \equiv (1 - p)^2 + (1 - (1 - p)^2)k^{-1}$, while nodes *not* on the planted path match with probability k^{-1}. Thus, the match of the planted path has score roughly $qd \pm O(\sqrt{qd})$, while nearly-disjoint paths have scores roughly $k^{-1}d \pm O(\sqrt{k^{-1}d})$. There are roughly m^d non-planted paths and they are close to disjoint, so standard large-deviation results for Gaussians suggest that the *highest* non-matching path has score roughly $k^{-1}d + \sqrt{k^{-1}d}\sqrt{2\log(m)d}$. Thus, we expect recovery roughly when

$$\left((1-p)^2 + (1-(1-p)^2)k^{-1}\right)d \gg \left(k^{-1} + \sqrt{2\log(m)k^{-1}}\right)d. \qquad (3)$$

In other words, we expect the planted path to have the highest score even when the branching factor m is fairly large and the alphabet size k is fairly small.

When Algorithm 1 is applied with weight function $w \equiv 1$, it returns the largest common sub-path of the two trees. It can be viewed as a reasonable "default" behaviour for a planted-path algorithm, but in practice one may want slightly different behaviour including (i) matching subtrees instead of paths, (ii) preference for shorter matches, (iii) reduce the number of "skipped" nodes, (iv) the "top-K" paths, rather than the single best match. Moreover, some matches may be more informative than others. A standard default choice is to "fit" the score function w to the data, setting

$$w(i,i) \propto \frac{2}{N(i) + 2}, \qquad (4)$$

where $N(i)$ is the number of times label i appears in the dataset (see *e.g.* Chap. 8 of [11] for an introduction to scoring rules of this nature).

3 Workflows

We expect Algorithm 1 to be a useful "building block" used as a step inside more complicated workflows. To show its power, we introduce below two that seem particularly relevant. More workflows will be discussed in the journal version of this paper.

3.1 Matching Workflow: Unsupervised Setting

One use of Algorithm 1 is to find a known, small template $\mathcal{H}$ inside of a large observed graph $\mathcal{G}$. In applications, we may not know a "good" template even if we have strong reason to believe a good template exists. In this situation, we may consider the problem of finding good matchings between two large graphs, then extracting candidate templates. The main algorithm here is a simple embedding and clustering algorithm based on the matching score.

As is typically the case for workflows that involve embedding and clustering, there is quite a bit of freedom to "switch out" components of our algorithm. As such, we give a somewhat-generic version of the workflow, Algorithm 2, with the following components being user-specified:

1. **Normalizer:** This is a function $\mathcal{N}$ from an n by n "similarity" matrix to an n by n "distance" matrix. To be concrete, in our experimental setting we use

the following sequence of calculations to find a distance matrix $D = \mathcal{N}(S)$ for a given symmetric similarity matrix S:

$$D[i,j] = \sqrt{1 - \frac{S[i,j]}{\sqrt{M[i]M[j]}}}, \qquad \text{where } M[i] \equiv \max_j S[i,j]. \qquad (5)$$

2. **Embedder:** This is a function $\mathcal{E}$ from an n by n "distance" matrix to an n by d "embedding" matrix. To be concrete, in our experimental setting we use UMAP (Uniform Manifold Approximation and Projection for Dimension Reduction) [13][3] with $d = 2$.
3. **Clusterer:** This is a function $\mathcal{C}$ from an n by d "embedding" matrix to a partition of $\{1, 2, \ldots, n\}$. To be concrete, in our experimental setting we use HDBSCAN (Hierarchical Density-Based Spatial Clustering of Applications with Noise) [12][4].

Algorithm 2. Unsupervised Clustering

Input: Trees $\{\mathcal{T}_i\}_{i=1}^N$ with labels $\{\phi_i\}_{i=1}^N$, weight function w, normalizer N, embedder E, clusterer C.

1 **for** $1 \leq i < j \leq N$: **do**
2   Find the best matching paths γ_i, γ_j and score S_{ij} by applying Algorithm 1 to input trees $\mathcal{T}_i, \mathcal{T}_j$, label functions ϕ_i, ϕ_j and weight function w.
3 **end for**
4 Compute the distance matrix $D = \mathcal{N}(S)$.
5 Compute the embedding $E = \mathcal{E}(D)$.
6 Compute and return the clustering $\mathcal{C}(E)$.

Given a cluster, we will typically want to learn the actual sequence that was planted in each tree in the cluster (either to allow us to cluster new data more efficiently, or because we believe the planted sequence is meaningful). Since Algorithm 2 involves computing a best-matched sequence in each pair of trees, this amounts to solving the well-known *multiple sequence alignment* (MSA) problem for this collection of best-matched sequences. The literature on this subject is too large to survey (see *e.g.* [3,4,9,10,17]); in this paper we use a simple likelihood-based approach similar to [17]. We will refer to this algorithm as MSA Algorithm.

3.2 Matching Workflow: Features for Classifiers

The simplest use for path matching is as a simple feature-augmentation technique. Consider an observed dataset of labelled trees $\{\mathcal{G}_i\}_{i=1}^N$ with associated feature maps $\{\phi_{\mathcal{G}_i}\}_{i=1}^N$. We then fix a collection of templates and feature maps $\{(\mathcal{T}_j, \phi_{\mathcal{T}_j})\}_{j=1}^M$. We can use these templates to generate auxiliary features $X_{i,j}$

---

[3] https://umap-learn.readthedocs.io/en/latest/.
[4] https://hdbscan.readthedocs.io/en/latest/.

by running Algorithm 1 with input $\mathcal{G}_i, \mathcal{T}_j, \phi_{\mathcal{G}_i}, \phi_{\mathcal{T}_j}$. These features can be used in any downstream task, *e.g.* as input for a classifier.

Of course, one must decide on where the pairs $\{(\mathcal{T}_j, \phi_{\mathcal{T}_j})\}_{j=1}^{M}$ come from. The simplest approach for selecting $\{(\mathcal{T}_j, \phi_{\mathcal{T}_j})\}_{j=1}^{M}$ is to randomly select paths that occur in your data - this amounts to using templates as random landmarks, as in *e.g.* [15].

4   Simple "Planted Path" Model

In this section, we describe two simple data-generating process for our planted-path problem: one for randomly planting a single fixed template in a single fixed tree (Algorithm 3), the other generating a collection of many trees that have hard-to-distinguish templates (Algorithm 4). We view these models as loosely analogous to using the stochastic block model (**SBM**) [5] or a family of Artificial Benchmarks for Community Detection (**ABCD**) [7] as a toy data-generating process for the clustering problem (see *e.g.* [8] for more examples of using random graphs in modelling and mining complex networks).

Before describing the process, we call attention to two pitfalls of using the **SBM** that we will try to avoid:

1. **Trivial Algorithms Work:** If one generates a large graph from the **SBM** with a small number of components and link probabilities chosen at random, then simply sorting nodes by degree will typically give a reasonably good clustering. We view this as a failure of the toy model, since we know that such "trivial" algorithms typically fail on realistic data.
2. **Generated Examples Too Dissimilar to Real Data:** The usual **SBM** provides unlabelled graphs. However, in practice, node features are extremely useful for clustering. We might expect clustering algorithms developed for the **SBM** to often generalize poorly.

We partially resolve the first problem by ensuring that, in Algorithm 4, the distribution of the *set* of labels $\{\phi(u)\}_{u \in \mathcal{T}}$ of each graph $\mathcal{T}$ is independent of the particular planted path. This ensures that trivial algorithms that do not use the tree structure, such as counting the number of times labels occur, must fail. We partially resolve the second problem by allowing a wide variety of underlying graph topologies and labels.

4.1   Labelling Trees with Planted Paths

Our basic model for planting a non-random sequence of features in a non-random tree is the following algorithm. We assume that we are given a tree $\mathcal{T}$ with a path Γ of length ℓ identified. Our goal is to plant a random subsequence of a given sequence of features of length k on a random part of the path Γ. Parameter p and function r model the length and, respectively, the distribution of a random subsequence. The remaining features are selected according to a distribution π.

Algorithm 3. Randomly Labeling Trees from Template

Input: Tree $\mathcal{T}$, connected path $\Gamma \equiv (\gamma_1, \ldots, \gamma_\ell) \subseteq \mathcal{T}$, distribution π on $\mathcal{S}$, sequence $a_1, \ldots, a_k \in \mathcal{S}$, rate $r : \mathcal{S} \mapsto (0, \infty)$, observation probability $0 < p \leq 1$.

1   Sample $N' \sim \mathrm{Bin}(\ell, p)$ and set $N = \min(N', k)$.

2   Sample set $S_1 \subseteq \{1, 2, \ldots, \ell\}$ of size N uniformly at random, and sample set $S_2 \subseteq \{1, 2, \ldots, k\}$ of size N with probability proportional to $\prod_{s \in S_2} r(s)$. Let $s_1(1) < \ldots < s_1(N)$, $s_2(1) < \ldots < s_2(N)$ be their elements in increasing order.

3   **for** $i \in \{1, 2, \ldots, N\}$ **do**

4   Set $\phi(\gamma_{s_1(i)}) = a_{s_2(i)}$.

5   **end for**

6   **for** $v \in \mathcal{T} \setminus \{\gamma_i\}_{i \in S_1}$ **do**

7   Randomly set $\phi(v) \sim \pi$.

8   **end for**

9   Return ϕ.

In a typical dataset, one might have a family of trees, each with one of many possible planted paths. Before describing an associated synthetic random model, we need some further notation. We say that μ is a *distribution on (tree, path) pairs* if a sample $(\mathcal{T}, \Gamma) \sim \mu$ consists of a directed tree $\mathcal{T}$ and a sequence $\Gamma = (\gamma_1, \ldots, \gamma_k) \subseteq \mathcal{T}$ that satisfies $\gamma_i <_\mathcal{T} \gamma_{i+1}$ for all $i \in \{1, 2, \ldots, k-1\}$.

For our experiments, we will use a simple model that fits our data reasonably well. Our model generates random trees $\mathcal{GW}(N, \lambda)$ with depth at most N and aims to produce "bushy" trees. Fix a positive integer N and a real number $\lambda > 1$. To generate a random tree, we will use a well-known Galton-Watson process in which the number of children of nodes are i.i.d. random variables that follow the Poisson distribution with mean λ.

Algorithm 4. Sampling from Toy Model

Input: Distribution μ on (tree, path) pairs, distribution π on $\mathcal{S}$, base sequence $a_1, \ldots, a_k \in \mathcal{S}$, rate $r : \mathcal{S} \mapsto (0, \infty)$, observation probability $0 < p \leq 1$, number of observations per class $n_1, \ldots, n_M$.

1   **for** $i \in \{1, 2, \ldots, M\}$: **do**

2   Sample permutation σ_i of $\{1, 2, \ldots, k\}$ uniformly at random.

3   **for** $j \in \{1, 2, \ldots, n_i\}$: **do**

4   Sample a tree and path $(\mathcal{T}_{ij}, \Gamma_{ij}) \sim \mu$.

5   Sample labels ϕ_{ij} according to Algorithm 3 with input tree $\mathcal{T}_{ij}$, path Γ_{ij}, distribution π, sequence $a_{\sigma_i(1)}, \ldots, a_{\sigma_i(k)}$, rate r, and observation probability p.

6   **end for**

7   **end for**

8   Return $\pi, \mathcal{T}_{ij}, \Gamma_{ij}, \phi_{ij}$.

In our basic model, Algorithm 4, all of the planted sequences $\{a_{\sigma_i(j)}\}_{j=1}^k$ are permutations of a single base sequence $\{a_j\}_{j=1}^k$. This has the advantage of ensuring that the distributions of sets of labels are completely independent of the class $i \in \{1, 2, \ldots, M\}$, ensuring that any algorithms based only on the sets of labels cannot distinguish between the classes. Of course, one could make a small tweak to this algorithm to allow any collection of planted sequences.

Algorithms 3 and 4 deal with two extreme cases: either both tree and planted path are fixed, or neither are fixed. Of course, it is reasonable to consider the intermediate situation for which the planted path is known and fixed while the trees are random. This corresponds to running Algorithm 4 with $M = 1$.

4.2 Sanity Check: Is This Problem Nontrivial but Tractable?

A natural question is: for our simulated data, how hard is it to distinguish a random tree *with* a planted path from a random tree *without* the path? Our goal is to check that this problem is nontrivial (in that simple approaches, such as label-counting, fail) and tractable (in that the problem would be fairly easy if you knew the path itself).

For this experiment, we use Algorithm 4 (with a simple base random tree generator) to generate 200 trees from $M = 2$ classes and observation probability $p = 0.75$, with a very small base alphabet $|\mathcal{S}| = 5$. Let us highlight the fact that a small alphabet with "bushy" trees ensures that there will be many "accidental" or "false positive" matches. Our scoring algorithm would look substantially better, even for much smaller values of p, if $|\mathcal{S}|$ were much larger.

The left-hand side of Fig. 2 shows a typical tree, with the candidate planted path highlighted (this is before subsampling at rate $p = 0.75$). The "similarity score" plotted on the right-hand side of Fig. 2 is computed by finding the similarity of each tree and the reference sequence a associated with class 1 using Algorithm 1. We then plot separately the similarities of all trees in class 1 ("within-cluster") and those in class 2 ("outside-cluster"). This experiment shows that our similarity score is very good at distinguishing between these two classes even when the planted path is both (i) a small fraction of the full "bushy" tree and (ii) sampled at a fairly small rate of $p = 0.75$.

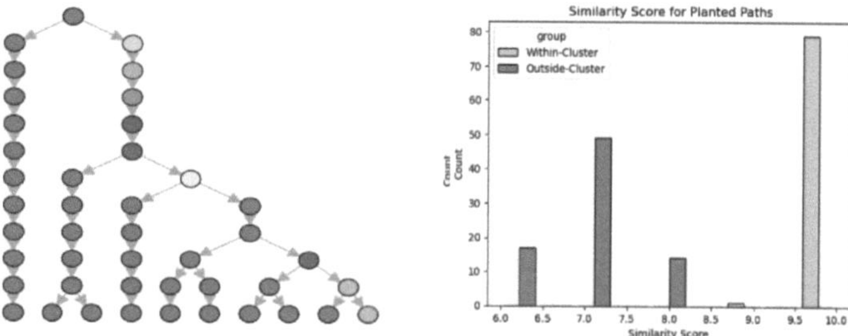

Fig. 2. Example of a typical tree (left) and histogram of similarity scores for the two classes (right).

Figure 3, on the other hand, shows that the histograms of symbol frequencies are indistinguishable. In these figures, for each tree, we compute the number of times a given symbol appears as a node label. We then plot the histogram of these counts for within- and outside-cluster trees.

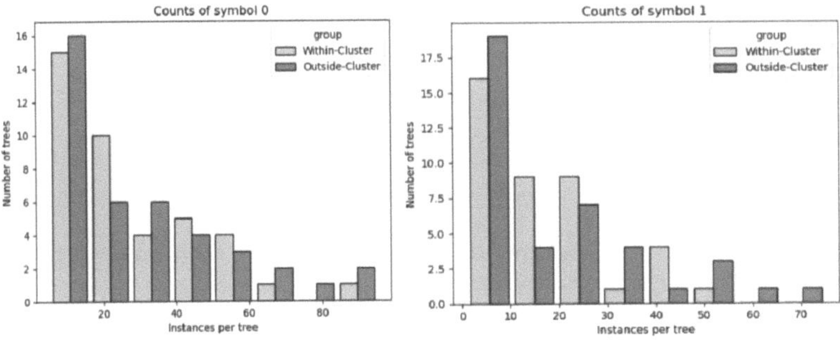

Fig. 3. Histograms of symbol frequencies.

4.3 Experiments on the Synthetic Model

We check that the unsupervised workflow proposed in Sect. 3.1 works reasonably well for this toy "planted path" model. In our first experiment, we generate labelled trees using Algorithm 4, with the main parameters being:

1. The base distribution on trees is $\mathcal{GW}(12, 1.8)$,
2. The base distribution π on the alphabet is uniform on $\mathcal{S} = \{A, B, C, D, E\}$,
3. The base sequence a is sampled from length-10 sequences using π,
4. The observation probability is $p = 0.9$, with rate r constant, and
5. The number of observations per class is $n_1 = \ldots = n_4 = 50$.

We then run Algorithm 2, with parameters as described in Sect. 3.1. The resulting embedding is presented in Fig. 4 (left). Additionally, we use our variant of the MSA Algorithm to extract exemplars, one for each family, and see whether they separate classes; see Fig. 4 (right). Even a quick visual inspection shows that both algorithms do fairly well, with the four ground-truth clusters well-separated and distance-to-exemplars much smaller for trees in the same class than trees in other classes.

Next, we consider essentially the same experiment, with four changes: we have $n_1 = \ldots = n_6 = 18$ observations per class and $p = 0.7$ (so that the base problem becomes slightly harder), we set $\pi(i) \approx (0.247, 0.247, 0.247, 0.247, 0.012)$ (that is, the symbol E is now rare), and we put two copies of E inside the base sequence a. We then run Algorithm 2 as above, but with two different choices for w: the "unweighted" choice $w \equiv 1$, and the "weighted" choice given in formula (4). The two resulting embeddings are displayed in Fig. 5. A quick visual inspection of Fig. 5 shows that a probability-weighted score function substantially improves separation of the clusters, as expected.

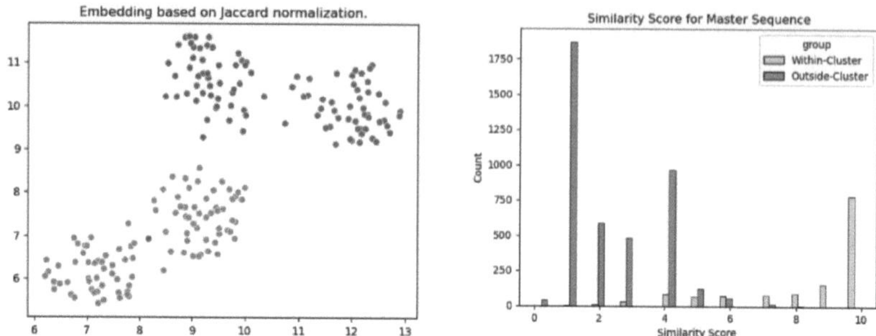

Fig. 4. Embedding of trees coloured by the "ground-truth" cluster label (left). Similarity score for the extracted exemplars (right).

5 Matching Paths in the ACME4 Dataset

ACME refers to cybersecurity datasets using the open-source Wintap telemetry collection tool developed by the Lawrence Livermore National Laboratory (LLNL). Designed specifically for cybersecurity research, **ACME** addresses the complexities of gathering and analyzing host-based data within large-scale Windows environments. Further documentation for both Wintap and **ACME** is available on-line[5].

We use the **ACME4** dataset, in particular, the tables defined in the standard view collection[6]. The **ACME4** data simulates a Windows business network with 10 workstations over a 2-week period and a variety of attacks (Living off the Land, Caldera, Metasploit, etc.). Simulated malicious actors are explicitly identified via "bad" username labels.

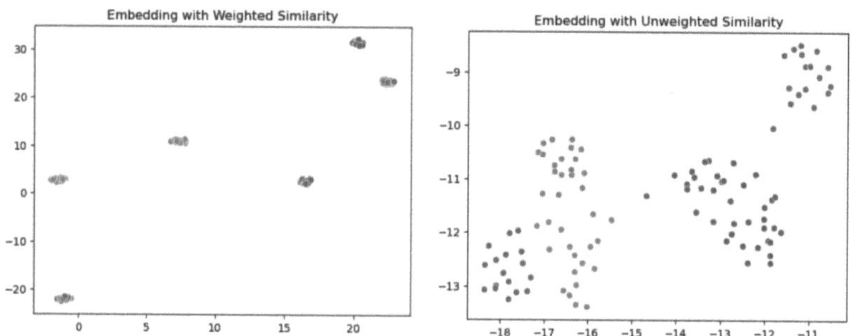

Fig. 5. Two embeddings of six classes of trees, coloured by the "ground-truth" cluster label: weighted (left) and "unweighted" (right).

---

[5] https://gdo168.llnl.gov.

[6] https://gdo168.llnl.gov/data/newdocs/datadict.

5.1   Process Trees

To build process trees, we extracted the following information from `process_uber_table` in the standard view collection for every process: the process identifier (PID) hash, its parent PID hash, the process name (often blank), and the user name (also often blank). From this data, we built a large directed graph where each edge is directed from parent to child. We then extracted the connected components, which correspond to the process trees. We obtained 1,111,277 trees, for a total of 2,884,171 nodes (PIDs) and 1,772,894 edges. The sizes of the trees range from 2 to 257,744, with the vast majority being of size 2.

Looking at every process tree, only 8 contain bad users. We show the smallest such tree in Fig. 6, where each node is indexed with a shortened and anonymized user name (such as "user88", "bad3" or "SYS") followed by the process name and separated with "::"; recall that both values can be blank.

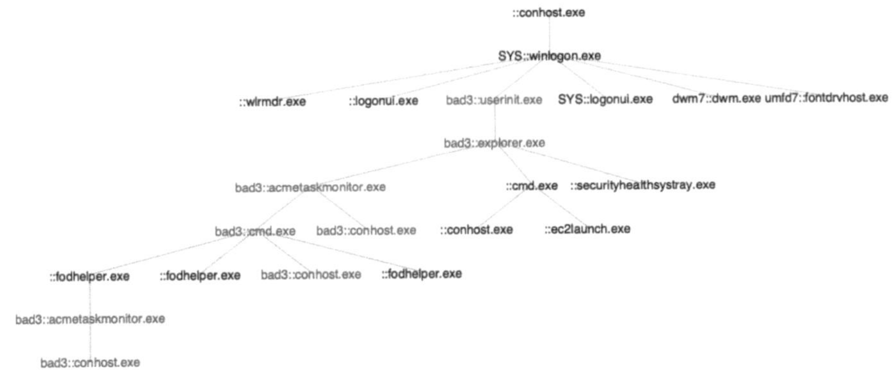

Fig. 6. Example of a process tree containing PIDs associated with a "bad". We use `short_username::process_name` as the node labels.

5.2   Matching Algorithm Experiments

In order to illustrate the use of matching algorithms such as Algorithm 1, we consider paths consisting of more than a single edge. Since the vast majority of process trees in the ACME4 dataset are of size 2, we build a larger set of sub-trees as follows.

For every process tree of size 3 to 10,000 (there are 210 of those), we consider every sub-tree of size 3 or more for which the root process name is not blank or labelled as "unknown". This yields 2,693 process sub-trees, with sizes ranging from 3 to 9,558 nodes, and depth from 2 to 15. Of those, 131 trees have at least one process associated with a "bad" user name. We selected one such tree as our "reference" tree, shown in Fig. 7(a). We then applied Algorithm 1, looking for matching path(s) with high scores in all other process trees. High scoring trees are shown in Fig. 7 using three different scoring functions to compare the node labels. In Fig. 7(b), we consider only the process names with a binary score (1 if

the processes are the same, and 0 otherwise). We see that we found a matching
path consisting of 6 nodes with the same process names as the path highlighted
in (a), thus with a total score of 6, but the path in (b) has blank usernames. Note
that we show all nodes not part of the matching path as simple dots for easier
visualization. In Fig. 7(c), we consider both the process names and usernames,
where all usernames starting with "user" or "bad" are considered as matching.
We use a binary function equal to 1 only if both the processes and usernames
match. We again found a matching path consisting of 6 nodes (with a total score
of 6) with the same process names as the path highlighted in (a), but the path
in (c) has username "user1" instead of "bad3" (and the two other usernames
also match, namely blank and SYS). In Fig. 7(d), we do as in (c), but we give
partial score for each matching part: 0.75 if the process names match, and 0.25
if the usernames match. We show a matching path with a total score of 5.75,
due to the fact that 3 of the 4 instances of "bad3" username in (a) correspond
to the non-blank username "user88" in (d), but the fourth username is blank. In
order to test if high-scoring matches really correspond to potentially malicious
activity, we considered a subset of the process trees with depth at least 4 and
size at most 100. From that set, we picked every tree with some "bad" user name
in turn, and applied Algorithm 1 for all other trees. From this experiment, we
found that given a randomly chosen "bad" tree, it is 11.6 times more likely that
a tree with maximum matching score also contains some "bad" user than if we
select this tree at random.

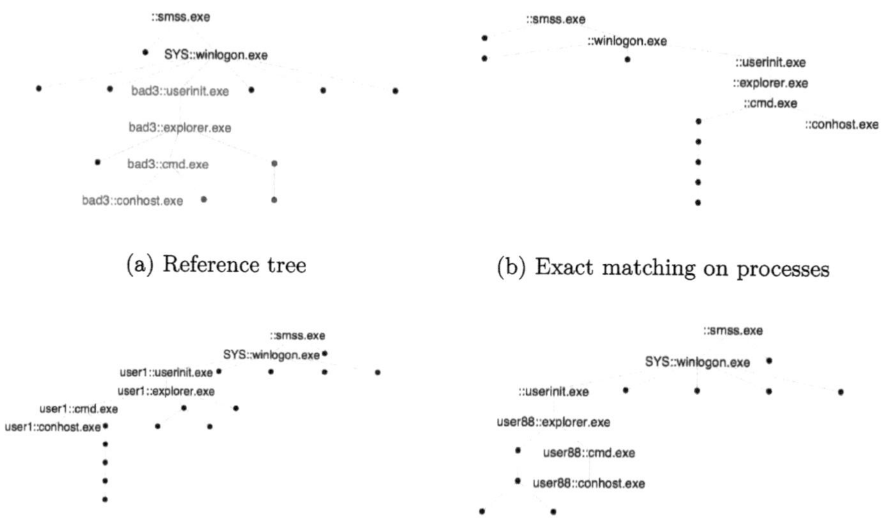

(a) Reference tree

(b) Exact matching on processes

(c) Exact matching on processes and users

(d) Soft matching on processes and users

Fig. 7. Results from Basic Matching Algorithm (Algorithm 1) using different similarity
scores for the node labels. Node labels are shown for the matching paths only.

The results shown in Fig. 7 can be seen as building blocks for more complex workflows, such as the ones described in Sect. 3.2. We illustrate two such workflows.

For the first illustration, we use a subset of the process trees we just described, keeping the ones with depth at least 4, and at most 100 nodes. There are 511 such trees, 38 of which contain at least one instance of a bad user, which we use as our templates T_j. For the other 473 process trees $\mathcal{G}_i$, we run our matching algorithm against every template and store the resulting scores (length of the longest common path) as features $X_{i,j}$. We then use those features to score the process trees as follows. For each $\mathcal{G}_i$, we compute $S_i = \sum_j \mathbf{1}_{\{X_{i,j} \geq 3\}}$, thus counting the number of templates having a matched path with 3 nodes or more. We plot the resulting score distribution in Fig. 8(a), where we see that a large number of trees have very low score, while a smaller group show a large score. This is an example where matching algorithm can be used as a filter, in this case selecting a subset of process trees having paths in common with several template trees. Looking more closely at the results, one clear difference is that trees $\mathcal{G}_i$ with score $S_i \geq 9$ tend to be larger than other trees, with respective median values of 22 and 6 nodes. Looking at the matching paths between templates and high scoring trees, the most frequent sequence of processes is `winlogon.exe-userinit.exe-explorer.exe-cmd.exe-conhost.exe` or a subset thereof, which coincides with the results shown in Fig. 7.

Four process names frequently found at the root of our process trees are: `bash.exe`, `cmd.exe`, `smss.exe` and `taskhostw.exe`. For the second illustration, we use the same 38 templates as before and we select a subset of 832 process trees $\mathcal{G}_i$, distinct from the templates, where the root process is one of those 4 processes. For each process tree, we run our matching algorithm against each template and store the resulting scores (length of the longest common path) as a 38-long feature vector $\{X_{i,j}, 0 \leq j \leq 37\}$. Randomly selecting half of the $\mathcal{G}_i$'s for training and the other half for testing, we trained a simple random forest classifier using the 4 process names as labels. The resulting confusion matrix is shown in Fig. 8(b), where we see that we can correctly classify most trees, with the exception of a small group of trees with root label `bash.exe` being classified as `cmd.exe`.

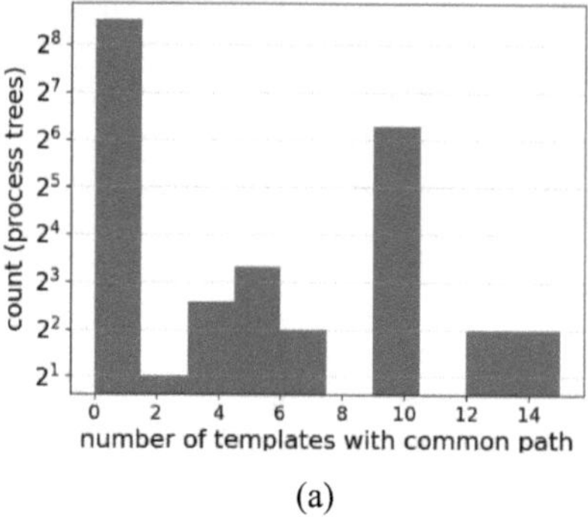
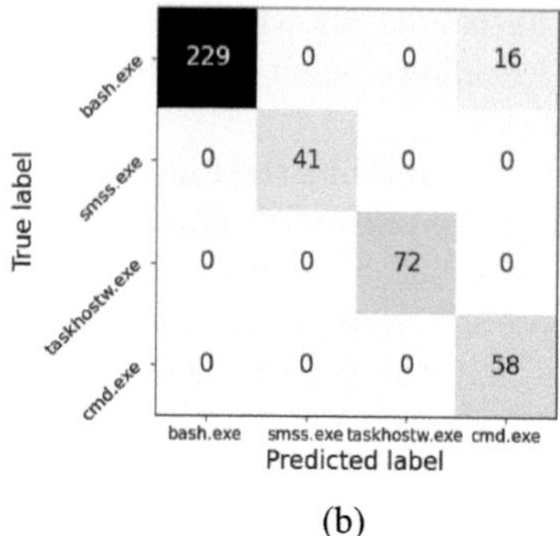

Fig. 8. (a) Counts of the number of process trees vs number of templates with a common path. (b) Confusion matrix for a random forest classifier using our matching algorithm to build the features.

acknowledgement. This research was supported by the Rogers Cybersecure Catalyst through the Catalyst Fellowship Program. The title was suggested by Gemini based on our introduction. It was justified like this. "It is my personal favorite. It twists the classic "needle in a haystack" idiom (which everyone in data science knows) to perfectly describe your specific contribution: you aren't looking for a single point, but a linear sequence (a thread)."

References

1. Anjum, M.M., Iqbal, S., Hamelin, B.: ANUBIS: a provenance graph-based framework for advanced persistent threat detection. In: Proceedings of the 37th ACM/SIGAPP Symposium on Applied Computing, SAC '22, pp. 1684–1693. ACM (2022)

2. Chen, Y., Jiaming, X.: Statistical-computational tradeoffs in planted problems and submatrix localization with a growing number of clusters and submatrices. J. Mach. Learn. Res. **17**(27), 1–57 (2016)

3. Do, C.B., Mahabhashyam, M.S.P., Brudno, M., Batzoglou, S.: ProbCons: probabilistic consistency-based multiple sequence alignment. Genome Res. **15**(2), 330–340 (2005)

4. Gotoh, O.: Multiple sequence alignment: algorithms and applications. Adv. Biophys. **36**, 159–206 (1999)

5. Holland, P.W., Laskey, K.B., Leinhardt, S.: Stochastic blockmodels: first steps. Soc. Netw. **5**(2), 109–137 (1983)

6. Jerrum, M.: Large cliques elude the metropolis process. Random Struct. Algorithms **3**(4), 347–360 (1992)

7. Kamiński, B., Prałat, P., Théberge, F.: Artificial benchmark for community detection (ABCD)–fast random graph model with community structure. Netw. Sci. **9**(2), 153–178 (2021)

8. Kamiński, B., Prałat, P., Théberge, F.: Mining Complex Networks. Chapman and Hall/CRC (2021)

9. Katoh, K., Standley, D.M.: MAFFT multiple sequence alignment software version 7: improvements in performance and usability. Mol. Biol. Evol. **30**(4), 772–780 (2013)
10. Lee, C.: Generating consensus sequences from partial order multiple sequence alignment graphs. Bioinformatics **9**(8), 999–1008 (2003)
11. Schütze, H., Manning, C.D., Raghavan, P.: Introduction to Information Retrieval. Cambridge University Press (2008)
12. McInnes, L., Healy, J., Astels, S.: HDBSCAN: hierarchical density based clustering. J. Open Source Softw. **2**(11), 205 (2017)
13. McInnes, L., Healy, J., Melville, J.: UMAP: uniform manifold approximation and projection for dimension reduction. arXiv preprint arXiv:1802.03426 (2018)
14. Microsoft Threat Intelligence: Seeing the big picture: deep learning-based fusion of behavior signals for threat detection (2020). Accessed 24 Dec 2025
15. Rahimi, A., Recht, B.: Random features for large-scale kernel machines. In: Platt, J., Koller, D., Singer, Y., Roweis, S. (eds.) Advances in Neural Information Processing Systems, vol. 20. Curran Associates, Inc. (2007)
16. Smith, T.D., Waterman, M.S.: Identification of common molecular subsequences. J. Mol. Biol. **147**, 195–197 (1981)
17. Thorne, J.L., Kishino, H., Felsenstein, J.: An evolutionary model for maximum likelihood alignment of DNA sequences. J. Mol. Evol. **33**(2), 114–124 (1991)
18. Usman, M., Jan, M.A., He, X., Chen, J.: A survey on representation learning efforts in cybersecurity domain. ACM Comput. Surv. **52**(6), 1–28 (2019)
19. Wu, Y., Xu, J.: Statistical problems with planted structures: Information-theoretical and computational limits. arXiv:1806.00118 (2018)

Algorithmic Aspects of the Clustering Regularity Lemma

Holden Pipp and Nicholas Sieger[(✉)]

University of California, San Diego, La Jolla, CA 92037, USA
{hpipp,nsieger}@ucsd.edu

Abstract. Clustering graphs, where the number of triangles is a constant fraction of the number of paths of length 2, appear frequently in real-world networks. Recently Chung showed that general clustering graphs are surprisingly structured in that they admit a regularity lemma; every clustering graph can be decomposed into a bounded number of random-like parts. We show that the regularity partition for clustering graphs can be computed in time $O(nM(\Delta(G)))$ where $M(n)$ is the complexity of $n \times n$ matrix multiplication and $\Delta(G)$ is the maximum degree of the graph. In particular, we show that it is possible to efficiently find subcommunities of a clustering graph which have robust clustering; even after removing many edges, the subcommunity has a large clustering coefficient. Our algorithm is parallelizable and can be implemented in NC^2. As a consequence of the proof, we show that sparse clustering graphs are closely related to dense quasirandom graphs.

Keywords: Clustering coefficient · regularity lemma · community detection

1   Introduction

Many real-world networks are *clustering graphs* where the presence of a common neighbor increases the likelihood that two nodes share an edge. Given the prevalence of such networks, with examples including social networks [1,2], biological networks [3], and semantic diagrams [4], the network and data science literature abounds with literature on statistical properties of clustering graphs [5,6] and algorithms for their analysis [7–9].The theoretical aspects of clustering graphs are less studied. To our knowledge, only the works of Gupta et al. [10] and Bourla et al. [7] tackles the graph-theoretic aspects of clustering graphs.

The recent works of Chung and the second author [11,12] showed that clustering graphs have a very nontrivial graph-theoretic structure: a regularity lemma. Regularity lemmas are patterned after the seminal work of Szemerédi's Regularity Lemma [13], which roughly states that *every* large dense graph has a vertex partition of bounded size such that almost all the graphs between the parts are random-like. Szemerédi's lemma plays a central role in the analysis of sets

© The Author(s), under exclusive license to Springer Nature Switzerland AG 2026
F. C. Graham et al. (Eds.): WAW 2026, LNCS 16630, pp. 82–96, 2026.
https://doi.org/10.1007/978-3-032-27193-8_6

without length-k arithmetic progressions [13,14], property testing in computer science [15], and extremal combinatorics [16,17].

One of the results of [12] gives a regularity lemma using the following notion of 'random-like.' A tripartite graph G is ϵ-*Triad Regular* if every subset of edges with sufficiently many 2-paths has a clustering coefficient within ϵ of the clustering coefficient of G. The regularity lemma of [12] informally states

Theorem 1. *[12, Theorem 4.3, informal] Fix $\epsilon > 0$. If a graph G has clustering coefficient γ, then the edge set can be partitioned into finitely many parts depending only on ϵ and γ, such that all but ϵ-many triangles in G are contained in the tripartite subgraphs of G which are ϵ-Triad Regular.*

We refer to Sect. 2 for the formal statement.

In this work, we show that the above regularity lemma for clustering graphs can be implemented algorithmically in polynomial time. More precisely, we show that a Triad-Regular partition can be found in time $O(nM(\Delta(G)))$ where $M(n)$ is the time to multiply two $n \times n$ matrices. Our work mirrors the result of Alon et al. [18] which gave a similar algorithmic implementation for Szemerédi's Regularity Lemma. However, the original Regularity Lemma and its implementation are only applicable for dense graphs, whereas our result applies to general clustering graphs which are frequently sparse [7]. Our proofs make use of the "local-to-global" methods from the study of High Dimensional Expanders [19–21] and reveal structural aspects of clustering graphs which may be of independent interest.

The paper is organized as follows. In Sect. 2 we give our definitions and formally state the particular version of the regularity lemma for clustering graphs in Theorem 2. In Sect. 3 we give two lemmas which together produce the main subroutine of our algorithm. In Sect. 4 we use the lemmas of Sect. 3 to show that Theorem 2 can be implemented in polynomial time in Theorem 3. In Sect. 5 we give some further directions and open problems.

2   Definitions and Preliminaries

Let A, B denote two subsets of the edge set of an undirected simple graph $G = (V, E)$. Note that A, B are not necessarily distinct. Let $G(A, B)$ denote the *tripartite subcover graph* with vertices $V_1 \sqcup V_2 \sqcup V_3$ where V_1, V_2, V_3 are copies of V. The edge set of $G(A, B)$ is the disjoint union $E_{12} \sqcup E_{23} \sqcup E_{13}$ where

$$E_{12} = \{uv \ : \ u \in V_1, v \in V_2, uv \in A\}$$
$$E_{23} = \{wv \ : \ v \in V_2, w \in V_3, vw \in B\}$$
$$E_{13} = \{uw \ : \ w \in V_3, u \in V_1, uw \in E\}.$$

Given $A, B \subseteq E$, we define the set of 2-paths in $G(A, B)$ with one edge in A and the other in B to be

$$P(A, B) = \{(u, v, w) \in V_1 \sqcup V_2 \sqcup V_3 : u \sim v \sim w \in G(A, B)\}$$

and the set of triangles with one edge in A and one edge in B to be

$$T(A, B) = \{(u, v, w) \in P(A, B) : w \sim u \in E\}.$$

The cardinalities of $P(A, B)$ and $T(A, B)$ will be denoted by $p(A, B)$ and $t(A, B)$ respectively.

The clustering coefficient of a tripartite subcover graph $G(A, B)$ is defined to be

$$\gamma^{(3)}(A, B) = \frac{t(A, B)}{p(A, B)}. \tag{1}$$

The clustering coefficient for a non-tripartite graph G, denoted $\gamma(G)$, is defined to be 3 times the number of unlabeled triangles in G divided by the number of unlabeled 2-paths. The clustering coefficient and tripartite clustering coefficient are related as follows. Since $t(E, E)$ counts every unlabled triangle 6 times and $p(E, E)$ counts each unlabeled 2-path 2 times and every degenerate path uvu 2 times, a short computation shows that:

$$\gamma^{(3)}(E, E) = \gamma(G)\left(1 - \frac{1}{\tilde{d}}\right) \tag{2}$$

where $\tilde{d}$ is the second-order average degree of G, i.e. $\tilde{d} = \frac{\sum_{v \in V} \deg_G(v)^2}{\sum_{v \in V} \deg_G(v)}$. As we will be working with graphs of large minimum degree (and therefore large $\tilde{d}$, Eq. 2 implies that $\gamma^{(3)}(G(E, E))$ and $\gamma(G)$ are essentially equivalent. We will focus on the tripartite clustering coefficient $\gamma^{(3)}(\cdot)$ henceforth.

With these concepts in hand, we can formally define Triad-Regularity as follows.

Definition 1. *Fix $\epsilon > 0$. For $A, B \subseteq E$, the tripartite graph $G(A, B)$ is said to be ϵ-Triad Regular if for any choice of $A' \subseteq A$ and $B' \subseteq B$ such that $p(A', B') \geq \epsilon p(A, B)$, we have*

$$|\gamma^{(3)}(A', B') - \gamma^{(3)}(A, B)| \leq \epsilon.$$

We say that a graph is ϵ-Triad-Irregular if it is not ϵ-Triad Regular.

The regularity lemma of [12] is then:

Theorem 2. *For any $\epsilon > 0$ and a graph $G = (V, E)$, let A, B be two subsets of E. Suppose that the tripartite graph $G(A, B)$ has clustering coefficient γ, then each of A, B can be partitioned into finitely many parts $S_1, S_2, \ldots, S_m$ for some m depending only on ϵ and γ, such that all but $\epsilon t(A, B)$ triangles in $G(A, B)$ are contained in the tripartite subgraphs $G(S_i, S_j)$ which are ϵ-Triad Regular.*

We will need the following definition to measure the size of a set of vertices in $G(A, B)$. The *pathweight* of a set $X \subset V_2$ with respect to the edge sets $A, B \subseteq E$ is

$$\mathrm{pwt}^{A,B}(X) = \sum_{v \in X} \deg_A(v) \deg_B(v)$$

and we write $\mathrm{pwt}(X)$ for $\mathrm{pwt}^{E,E}(X)$.

3   Deviation and Triad Regularity

Our goal is to show that, given $A, B \subseteq E$, we can either verify that (A, B) is ϵ-Triad Regular or find witness sets $A' \subseteq A$ and $B' \subseteq B$ such that the clustering coefficient $\gamma(A', B')$ differs from the clustering coefficient $\gamma(A, B)$ by a small δ. Ideally, we would find a witness (A', B') of ϵ-Triad Irregularity, i.e. $|\gamma(A', B') - \gamma(A, B)| > \epsilon$ and $p(A', B') > \epsilon p(A, B)$. However, Alon et al. [18] show that finding a witness of ϵ-irregularity (in the setting of the original regularity lemma) is NP-Hard, and we suspect an analogous result holds in our setting. So instead, we will simply try to find a witness of δ-Triad Irregularity for a smaller $\delta = \delta(\epsilon)$.

Facing an analogous problem, Alon et al. introduce the *deviation* of a set of vertices. Their deviation property states that the total number of common neighbors is far from its expected value in the random graph $G(n, p)$. Unlike ϵ-irregularity, the deviation can be computed in polynomial time via matrix multiplication. Given a set with large deviation, Alon et al. show that two implications. First, they show that any set with large deviation must contain a witness of δ-irregularity for $\delta \approx \epsilon^4$. Furthermore, the witness can be found in polynomial time. Second, they show small deviation implies ϵ-regularity. In fact the two graph properties are part of a larger class of polynomially equivalent[1] *quasirandom* graph properties found in [22].

Now we turn to how we will apply Alon et al.'s general proof strategy to the setting of ϵ-Triad Regularity. Whereas the original regularity lemma assumes that the whole graph is dense, we only assume that many 2-paths are contained in a triangle. As a result, the neighborhood graphs $G[N(v)]$ should be largely dense, an intuition formalized in [10]. One may suspect that if a clustering graph satisfies the stronger condition of ϵ-Triad Regularity, perhaps the neighborhood graphs are even quasirandom graphs themselves.

Our key conceptual contribution formalizes the above intuition and confirms that most of the neighborhood graphs of ϵ-Triad-Regular graphs are quasirandom graphs. We define an version of deviation for clustering graphs (found in Definition 3), which is simply the set of vertices whose neighborhood graphs have large deviation in the sense of Alon et al. We then show that our new notion of deviation is polynomially equivalent to ϵ-Triad Regularity in Lemmas 1 and 2. Finally, we show that a set with large deviation must contain a δ-Triad Irregular pair (A', B') and that this pair can be found in polynomial time. We note that the polynomial equivalence of ϵ-Triad Regularity and small local deviations is part of a larger quasirandom equivalence theorem which will appear in a forthcoming work [23].

Now to the details. Given $A, B \subseteq E$, a vertex $v \in V_2$ and vertices $u_1, u_2 \in V_1$ such that $u_1 v, u_2 v \in A$, the *specific deviation* $\sigma_v^{A,B}(u_1, u_2)$ is

$$\sigma_v^{A,B}(u_1, u_2) := |N(u_1) \cap N(u_2) \cap N_B(v)| - \gamma^2 \deg_B(v).$$

---

[1] A graph property $P(\delta)$ *polynomially implies* property $Q(\epsilon)$ if there is a polynomial $p(x)$ such that for any $\delta < p(\epsilon)$, $P(\delta) \implies Q(\epsilon)$. Two graph properties are *polynomially equivalent* if they polynomially imply each other.

Our key definition is the following:

Definition 2. *Given $A, B \subseteq E$ and a vertex $v \in V_2$, the local deviation is*

$$\sigma_v(A, B) := \sum_{\substack{u_1, u_2 \in V_1 \\ u_1 v \in A \\ u_2 v \in A}} \sigma_v^{A,B}(u_1, u_2)$$

Our goal is to show that if the local deviations of a tripartite graph $G(A, B)$ are greater than some function of ϵ, then $G(A, B)$ cannot be ϵ-Triad Regular. We will need the following definitions.

The *deviation set* of A, B is

$$\mathrm{Dev}^{A,B}(\delta) := \left\{ v \in V_2 : \sigma_v(A, B) > \delta \deg_A(v)^2 \deg_B(v) \right\}.$$

The *pathweight* of a set $X \subseteq V_2$ with respect to $A \subseteq E_{12}, B \subseteq E_{23}$ is

$$\mathrm{pwt}^{A,B}(X) := \sum_{v \in X} \deg_A(v) \deg_B(v).$$

Definition 3. *Edge sets $A, B \subseteq E$ have (δ_1, δ_2)-small deviations if for every $A' \subseteq A$ and $B' \subseteq B$, the deviation set of A', B' has at most a δ_2-fraction of all paths using A, B, i.e. $\mathrm{pwt}^{A,B}(\mathrm{Dev}^{A',B'}(\delta_1)) < \delta_2 p(A, B)$.*

We will also need to track local degrees.

Definition 4. *For edge sets $A, B \subseteq E$, a vertex $v \in V_2$ is (ϵ, δ)-triangle irregular with respect to (A, B) if there are at least $\delta \deg_A(v)$ vertices $u \in V_1$ such that*

$$|\,|N(u) \cap N_B(v)| - \gamma \deg_B(v)\,| > \epsilon \deg_B(v)$$

First we show that if A, B have small deviations, they admit a sufficiently large quantity of triangle-regular vertices in V_2.

Lemma 1. *Fix $A, B \subseteq E$, and $\epsilon < \frac{1}{16}$. Let $R \subseteq V_2$ be the set of vertices which are $(\frac{\epsilon^5}{90}, \frac{\epsilon^{5/2}}{9})$-triangle irregular. If $\mathrm{pwt}^{A,B}(R) < \frac{2\epsilon^{5/2}}{5} p(A, B)$ and $G(A, B)$ has $(\frac{\epsilon^2}{9}, \frac{2\epsilon^2}{5})$-small deviations, then (A, B) is ϵ-Triad Regular.*

Proof. Fix $X \subseteq A$ and $Y \subseteq B$ such that $p(X, Y) > \epsilon p(A, B)$.

We first remove all irregular edges from X, by isolating all edges which are not triangle-regular.

Let $\partial R = \{uv \in A; v \in R\}$ and let

$$Z = \left\{ uv \in X : v \notin R \wedge |\,|N(u) \cap N_B(v)| - \gamma \deg_Y(v)\,| > \frac{\epsilon^5}{90} \deg_Y(v) \right\}.$$

Let $X_0 = X \setminus (Z \cup \partial R)$ and $X' = X \cap (Z \cup \partial R)$. The triangle inequality then implies that

$$\begin{aligned}
|t(X, Y) - \gamma p(X, Y)| &\leq |t(X_0, Y) - \gamma p(X_0, Y)| + |t(X', Y) - \gamma p(X', Y)| \\
&\leq |t(X_0, Y) - \gamma p(X_0, Y)| + p(X', Y) \\
&= |t(X_0, Y) - \gamma p(X_0, Y)| + p(Z, Y) + p(\partial R, Y).
\end{aligned}$$

For the term $P(Z, Y)$, we note Z only contains edges incident to $(\frac{\epsilon^5}{90}, \frac{\epsilon^{5/2}}{9})$-triangle regular vertices. Therefore, we see that $\deg_Z(v) < \frac{\epsilon^{5/2}}{9} \deg_X(v)$ for each $v \in V_2$. Hence,

$$p(Z, Y) = \sum_{v \in V_2 \setminus R} \deg_Z(v) \deg_Y(v) \le \sum_{v \in V_2} \frac{\epsilon^{5/2}}{9} \deg_X(v) \deg_Y(v) = \frac{\epsilon^{5/2}}{9} p(X, Y).$$
$$(3)$$

For the remaining term, we note that $\deg_{\partial R}(v) = \deg_X(v)$ and thus

$$p(\partial R, Y) = \mathrm{pwt}^{X,Y}(R) \le \mathrm{pwt}^{A,B}(R) \le \frac{2\epsilon^{5/2}}{5} p(A, B) \qquad (4)$$

since $\mathrm{pwt}^{A,B}(R) \le \frac{\epsilon^{5/2}}{9} p(A, B)$ by assumption.

Thus we focus on $|t(X_0, Y) - \gamma p(X_0, Y)|$ and additionally isolate the set of vertices with large deviation. Define X_1, Y_1 by removing all edges in respectively X_0 and Y incident to the deviation set $\mathrm{Dev}^{X_0, Y}(\frac{\epsilon^2}{9})$, and let X_2, Y_2 be the edges incident to $\mathrm{Dev}^{X_0, Y}(\frac{\epsilon^2}{9})$. We then have

$$|t(X_0, Y) - \gamma p(X_0, Y)| = |t(X_1, Y_1) - \gamma p(X_1, Y_1)| + |t(X_2, Y_2) - \gamma p(X_2, Y_2)|$$

$$\le |t(X_1, Y_1) - \gamma p(X_1, Y_1)| + \mathrm{pwt}^{A,B}\left(\mathrm{Dev}^{X_0, Y}\left(\frac{\epsilon^2}{9}\right)\right)$$

$$\le |t(X_1, Y_1) - \gamma p(X_1, Y_1)| + \frac{2\epsilon^2}{5} p(A, B)$$

where we use the assumption that $\mathrm{pwt}^{A,B}\left(\mathrm{Dev}^{X_0, Y}\left(\frac{\epsilon^2}{9}\right)\right) \le \frac{2\epsilon^2}{5} p(A, B)$. We will write $[uw]$ as shorthand for the indicator $[uw \in E_{13}]$. We then have

$$|t(X_1, Y_1) - \gamma p(X_1, Y_1)| = \left| \sum_{\substack{v \in V_2}} \sum_{\substack{u \in V_1 \\ uv \in X_1}} \sum_{\substack{w \in V_3 \\ wv \in Y_1}} ([uw] - \gamma) \right|$$

$$\le \sum_{v \in V_2} \left| \sum_{\substack{w \in V_3 \\ wv \in Y_1}} \sum_{\substack{u \in V_1 \\ uv \in X_1}} ([uw] - \gamma) \right| \qquad (5)$$

We then focus on the inner absolute value, aiming to get a term similar to our definition of specific deviation. By Cauchy-Schwarz, we have

$$\left| \sum_{\substack{w \in V_3 \\ wv \in Y_1}} \sum_{\substack{u \in V_1 \\ uv \in X_1}} ([uw] - \gamma) \right|^2 \le \deg_{Y_1}(v) \left(\sum_{\substack{w \in V_3 \\ wv \in Y_1}} \left(\sum_{\substack{u \in V_1 \\ uv \in X_1}} ([uw] - \gamma) \right)^2 \right) \qquad (6)$$

Let $K(v) = \sum_{\substack{w \in V_3 \\ wv \in Y_1}} \left(\sum_{\substack{u \in V_1 \\ uv \in X_1}} ([uw] - \gamma) \right)^2$ from Eq. 6. This is exactly the first term in σ_v. Expanding the inner square, we see that

$$K(v) = \sum_{\substack{w \in V_3 \\ wv \in Y_1}} \sum_{\substack{u_1, u_2 \\ u_1 v \in X_1 \\ u_2 v \in X_1}} ([u_1 w][u_2 w]) - \gamma[u_1 w] - \gamma[u_2 w] + \gamma^2$$

$$= \left(\sum_{\substack{u_1, u_2 \\ u_1 v \in X_1 \\ u_2 v \in X_1}} |N(u_1) \cap N(u_2) \cap N_{Y_1}(v)| \right)$$

$$- 2\gamma \deg_{X_1}(v) e_v(X_1, Y_1) + \gamma^2 \deg_{X_1}(v)^2 \deg_{Y_1}(v)$$

$$= \sigma_v(X_1, Y_1) - 2\gamma \deg_{X_1}(v) e_v(X_1, Y_1) + 2\gamma^2 \deg_{X_1}(v)^2 \deg_{Y_1}(v)$$

where $e_v(X_1, Y_1) = \{uw \in E_{13} : uv \in X_1, wv \in Y_1\}$. By our definition of X_1, we have $||N(u) \cap N_{Y_1}(v)| - \gamma \deg_{Y_1}(v)| < \frac{\epsilon^5}{90} \deg_{Y_1}(v)$ for all $uv \in X_1$. It follows that

$$|e_v(X_1, Y_1) - \gamma \deg_{X_1}(v) \deg_{Y_1}(v)| \leq \sum_{u:uv \in X_1} |\deg_{Y_1}(u) - \gamma \deg_{Y_1}(v)|$$

$$\leq \sum_{u:uv \in X_1} \frac{\epsilon^5}{90} \deg_{Y_1}(v)$$

$$= \frac{\epsilon^5}{90} \deg_{X_1}(v) \deg_{Y_1}(v)$$

and therefore

$$K(v) \leq \sigma_v(X_1, Y_1) + 2\frac{\epsilon^5}{90}\gamma \deg_{X_1}(v)^2 \deg_{Y_1}(v)$$

$$\leq \left(\frac{\epsilon^2}{9} + 2\frac{\epsilon^5}{90}\gamma \right) \deg_{X_1}(v)^2 \deg_{Y_1}(v) \tag{7}$$

where we use the fact that v is not in the deviation set $\mathrm{Dev}^{X_0, Y}(\frac{\epsilon^2}{9})$ and thus $\sigma_v(X_1, Y_1) = \sigma_v(X_0, Y) < \frac{\epsilon^2}{9} \deg_{X_1}(v)^2 \deg_{Y_1}(v)$. Thus for every $v \notin \mathrm{Dev}^{X_0, Y}(\frac{\epsilon^2}{9})$, Eqs. (6) and (7) imply that

$$\left| \sum_{\substack{w \in V_3 \\ wv \in Y_1}} \sum_{\substack{u \in V_1 \\ uv \in X_1}} ([uw] - \gamma) \right|^2 \leq \left(\frac{\epsilon^2}{9} + 2\frac{\epsilon^5}{90}\gamma \right) \deg_{X_1}(v)^2 \deg_{Y_1}(v)^2$$

and hence by Eq. (5) and the fact that $\gamma < 1$,

$$|t(X_1, Y_1) - \gamma p(X_1, Y_1)| \leq \sqrt{\frac{\epsilon^2}{9} + 2\frac{\epsilon^5}{90}} \sum_{v \in V_2} \deg_{X_1}(v) \deg_{Y_1}(v)$$

$$= \sqrt{\frac{\epsilon^2}{9} + 2\frac{\epsilon^5}{90}} p(X_1, Y_1) \tag{8}$$

By combining Eqs. (3), (4), (8), we conclude

$$|t(X,Y) - \gamma p(X,Y)| \leq \left(\sqrt{\frac{\epsilon^2}{9} + 2\frac{\epsilon^5}{90}} + \frac{\epsilon^{5/2}}{9}\right) p(X_1, Y_1) + \left(\frac{2\epsilon^{5/2}}{5} + \frac{2\epsilon^2}{5}\right) p(A,B)$$

$$\leq \left(\frac{\epsilon}{4}\sqrt{1 + \frac{\epsilon^3}{5}} + \frac{\epsilon^{5/2}}{16} + \epsilon^{3/2} + \frac{\epsilon}{2}\right) p(X,Y)$$

$$\leq \left(\frac{\epsilon}{4}\left(1 + \frac{2^{-13}}{5}\right) + \frac{\epsilon}{576} + \frac{\epsilon}{2}\right) p(X,Y)$$

$$\leq \epsilon p(X,Y)$$

where we use the facts that $p(X,Y) \geq \epsilon p(A,B)$ and that $\epsilon < \frac{1}{16}$. We conclude that (X,Y) is ϵ-Triad Regular as desired. $\qquad\square$

Now we can show a converse, that large deviation detects a Triad Irregular pair.

Lemma 2. *Assume $\epsilon < \frac{1}{16}$. Fix $A, B \subseteq E$. Suppose that (A,B) has $(\frac{\epsilon^2}{9}, \frac{2\epsilon^2}{5})$-large deviations. Then one of the following must hold:*

1. *$t(A,B) < \epsilon p(A,B)$.*
2. *There is a set $X \subseteq V_2$ such that every vertex $v \in X$ is $(\frac{\epsilon^5}{90}, \frac{\epsilon^{5/2}}{9})$-triangle irregular and $\mathrm{pwt}^{A,B}(X) > \frac{2\epsilon^{5/2}}{5} p(A,B)$.*
3. *There are sets of edges $A' \subseteq A$ and $B' \subseteq B$ such that*

$$|\gamma(A', B') - \gamma(A,B)| > \frac{\epsilon^5}{90}$$

and $p(A', B') > \frac{\epsilon^5}{90} p(A,B)$.

Furthermore, there is an $O(nM(\Delta(G)))$-time NC^2 algorithm which produces the pair (A', B') in step 3.

Proof. We assume that neither (1) nor (2) hold, and then show that (3) must hold. Our first step is to show that there is a large subset of the deviation set $Dev^{A,B}(\frac{\epsilon^2}{9})$ which is $(\frac{\epsilon^5}{90}, \frac{\epsilon^{5/2}}{9})$-triangle regular. As (2) does not hold, the set of $(\frac{\epsilon^5}{90}, \frac{\epsilon^{5/2}}{9})$-triangle irregular vertices has $\mathrm{pwt}^{A,B}(X) \leq \frac{2\epsilon^{5/2}}{5} p(A,B)$. Thus removing all $(\frac{\epsilon^5}{90}, \frac{\epsilon^{5/2}}{9})$-triangle irregular vertices from $Dev^{A,B}(\frac{\epsilon^2}{9})$, removes at most a $\frac{2\epsilon^{5/2}}{5}$ fraction of the pathweight. Hence there exists a subset $Y \subseteq Dev^{A,B}(\frac{\epsilon^2}{9})$ with $\mathrm{pwt}^{A,B}(Y) \geq (\frac{2\epsilon^2}{5} - \frac{2\epsilon^{5/2}}{5}) p(A,B)$ such that for every vertex $v \in Y$, there is a corresponding subset $U_v \subseteq N_A(v)$ with $|U_v| \geq (1 - \frac{\epsilon^{5/2}}{9}) \deg_A(v)$ and every $u \in U_v$ satisfies

$$||N(u) \cap N_b(v)| - \gamma \deg_B(v)| \leq \frac{\epsilon^5}{90} \deg_B(v). \tag{9}$$

We note that

$$\mathrm{pwt}^{A,B}(Y) \geq \frac{3}{10}\epsilon^2 p(A,B) \tag{10}$$

as $\epsilon < \frac{1}{16}$.

Fix a vertex $v \in Y$, and consider U_v. Let $u_v^* \in U_v$ be a vertex which maximizes the sum of specific deviations with itself and the rest of U_v, i.e. u_v^* maximizes the quantity $\sum_{u_2 \in N_A(v)} \sigma_v^{A,B}(u_v^*, u_2)$. We wish to lower bound $\sum_{u_2 \in N_A(v)} \sigma_v^{A,B}(u_v^*, u_2)$. We first note that

$$\sum_{u_1 \in U_v} \sum_{u_2 \in N_A(v)} \sigma_v^{A,B}(u_1, u_2) \leq \deg_A(v) \sum_{u_2 \in N_A(v)} \sigma_v^{A,B}(u_v^*, u_2) \qquad (11)$$

since u_v^* maximizes the inner sum. On the other hand,

$$\sum_{u_1 \in U_v} \sum_{u_2 \in N_A(v)} \sigma_v^{A,B}(u_1, u_2) \geq \sigma_v(A, B) - \sum_{u_1 \in N_A(v)} \sum_{u_2 \in N_A(v) \setminus U_v} \sigma_v^{A,B}(u_1, u_2)$$

Since v is $Y \subseteq Dev^{A,B}(\frac{\epsilon^2}{9})$, $\sigma_v(A, B) \geq \frac{\epsilon^2}{9} \deg_A(v)^2 \deg_B(v)$. Since $\sigma_v^{A,B}(u_1, u_2) \leq \deg_B(v)$ for any u_1, u_2,

$$\sum_{\substack{u_1 \in N_A(v) \\ u_2 \in N_A(v) \setminus U_v}} \sigma_v^{A,B}(u_1, u_2) \leq \sum_{u_1 \in N_A(v)} \sum_{u_2 \in N_A(v) \setminus U_v} \deg_B(v) \leq \frac{\epsilon^{5/2}}{9} \deg_A(v)^2 \deg_B(v)$$

where we use the fact that $|N_A(v) \setminus U_v| < \frac{\epsilon^{5/2}}{9} \deg_A(v)$ for $v \in Y$. It follows that

$$\sum_{u_1 \in U_v} \sum_{u_2 \in N_A(v)} \sigma_v^{A,B}(u_1, u_2) \geq \left(\frac{\epsilon^2}{9} - \frac{\epsilon^{5/2}}{9} \right) \deg_A(v)^2 \deg_B(v) \qquad (12)$$

Combining Eqs. (11) and (12) we see that

$$\sum_{u_2 \in N_A(v)} \sigma_v^{A,B}(u_v^*, u_2) \geq \left(\frac{\epsilon^2}{9} - \frac{\epsilon^{5/2}}{9} \right) \deg_A(v) \deg_B(v). \qquad (13)$$

Now we show that there is a large set of vertices in $U_v(v)$ whose specific deviation with u_v^* is bounded below. Let $L_v \subseteq N_A(v)$ be the set of vertices u_2 such that $\sigma_v^{A,B}(u_v^*, u_2) > \epsilon^5 \deg_B(v)$. Since $\sigma_v^{A,B}(u_v^*, u_2) \leq \deg_B(v)$ for any $u_2 \in L_v$, we have

$$\sum_{u_2 \in N_A(v)} \sigma_v^{A,B}(u_v^*, u_2) \leq |L_v| \deg_B(v) + \epsilon^5(|U_v| - |L_v|) \deg_B(v).$$

As $|U_v| \leq \deg_A(v)$, Eq. (13) implies that

$$\begin{aligned}
|L_v| &\geq \frac{\frac{\epsilon^2}{9} - \frac{\epsilon^{5/2}}{9} - \epsilon^5}{1 - \epsilon^5} \deg_A(v) \\
&\geq \left(\frac{\epsilon^2}{9} - \frac{\epsilon^{5/2}}{9} - \epsilon^5 \right) \deg_A(v) \\
&\geq \frac{\epsilon^2}{9} \left(1 - \frac{1}{4} - \frac{1}{256} \right) \deg_A(v) \\
&= \frac{7\epsilon^2}{90} \deg_A(v), \qquad (14)
\end{aligned}$$

where we use the fact that $\epsilon < \frac{1}{16}$.

Let $A_v = \{u_2 v : u_2 \in L_v\}$. Similarly, define $B_v = \{vw : w \in N(u_v^*) \cap N_B(v)\}$. Then let

$$A' = \bigcup_{v \in Y} A_v \qquad B' = \bigcup_{v \in Y} B_v.$$

By Eq. (14), for any $v \in \mathrm{Dev}^{A,B}(\frac{\epsilon^2}{9})$. Hence,

$$\deg_{A'}(v) \geq \frac{7\epsilon^2}{90} \deg_A(v). \tag{15}$$

Since $u_v^* \in U_v$, Eq. (9) implies that $|\deg_{B'}(v) - \gamma \deg_B(v)| < \frac{\epsilon^5}{90} \deg_B(v)$.

We claim that $\gamma(A', B') - \gamma > \frac{\epsilon^5}{90}$ and $p(A', B') > \frac{\epsilon^5}{90} p(A, B)$. We first lower bound $p(A, B)$. By Eq. (15)

$$p(A', B') = \sum_{v \in Y} \deg_{A'}(v) \deg_{B'}(v) \geq \frac{7\epsilon^2}{90} \left(\gamma - \frac{\epsilon^5}{90} \right) \mathrm{pwt}^{A,B}(Y) \tag{16}$$

Since $\mathrm{pwt}^{A,B}(Y) \geq (\frac{2\epsilon^2}{5} - \frac{2\epsilon^{5/2}}{5}) p(A, B)$,

$$p(A', B') \geq \frac{7\epsilon^2}{90} \left(\gamma - \frac{\epsilon^5}{90} \right) \left(\frac{2\epsilon^2}{5} - \frac{2\epsilon^{5/2}}{5} \right) p(A, B) \tag{17}$$

$$\geq \frac{7\epsilon^2}{90} \epsilon \left(1 - \frac{1}{90 \cdot 16^4} \right) \frac{3\epsilon^2}{10} \tag{18}$$

$$\geq \frac{\epsilon^5}{90} p(A, B) \tag{19}$$

where we use the facts that $\gamma > \epsilon$ and $\epsilon < \frac{1}{16}$.

Now we show that $\gamma(A', B')$ is bounded away from $\gamma(A, B)$. We first note that

$$\sigma_v^{A,B}(u_v^*, u) = |N(u_v^*) \cap N(u) \cap N_B(v)| - \gamma^2 \deg_B(v) = |N(u) \cap N_{B'}(v)| - \gamma^2 \deg_B(v).$$

Since $|\deg_{B'}(v) - \gamma \deg_B(v)| < \frac{\epsilon^5}{90} \deg_B(v)$ by Eq. (9), it follows that

$$|N(u) \cap N_{B'}(v)| - \gamma \deg_{B'}(v) \geq |N(u) \cap N_{B'}(v)| - \gamma^2 \deg_B(v) - \gamma \frac{\epsilon^5}{90} \deg_B(v)$$

$$= \sigma_v^{A,B}(u_v^*, u) - \gamma \frac{\epsilon^5}{90} \deg_B(v).$$

Hence

$$|t(A', B') - \gamma p(A', B')| = \left| \sum_{v \in Y} \sum_{u \in N_{A'}(v)} |N(u) \cap N_{B'}(v)| - \gamma \deg_{B'}(v) \right|$$

$$\geq \left| \sum_{v \in Y} \sum_{u \in L_v} \sigma_v^{A,B}(u_v^*, u) - \gamma \frac{\epsilon^5}{90} \deg_B(v) \right|$$

By the definition of A_v and L_v, we have $\sigma_v^{A,B}(u_v^*, u) \geq \epsilon^5 \deg_B(v)$ for each $u \in L_v$. Hence,

$$|t(A', B') - \gamma p(A', B')| \geq \left| \sum_{v \in Y} \deg_{A'}(v)(\epsilon^5 - \gamma \frac{\epsilon^5}{90}) \deg_B(v) \right|$$

Again using Eq. (9) to bound $\deg_{B'}(v)$, we conclude that

$$|t(A', B') - \gamma p(A', B')| \geq \frac{\epsilon^5 - \gamma \frac{\epsilon^5}{90}}{\gamma - \frac{\epsilon^5}{90}} \sum_{v \in Y} \deg_{A'}(v) \deg_{B'}(v) \geq \frac{\epsilon^5}{90} p(A', B')$$

since $\gamma < 1$ and $\frac{1 - \frac{\gamma}{90}}{\gamma - \frac{\epsilon^5}{90}} \geq 1 - \frac{1}{90} \geq \frac{1}{90}$. We note that the specific deviation of a vertex v can be computed in time $O(M(\deg(v)))$ by squaring the bipartite adjacency matrix of the graph induced by $N_A(v)$ and $N_B(v)$. Given the square of the bipartite adjacency matrix, we can compute the sets U_v, the vertex u_v^*, and the sets A_v and B_v in time $O(\deg(v)^2)$. All other steps involve summing a quantity over all vertices, and can be done in time $O(n)$. Hence the overall running time is $O(nM(\Delta(G)))$, and the parallelization is clear. $\quad\square$

Putting Lemmas 1 and 2 together gives us an subroutine which either certifies that a pair (A, B) is ϵ-Triad Regular for a fixed $\epsilon > 0$ or gives a witness set of δ-Triad Irregularity for some δ which depends only on ϵ.

Corollary 1. *For $\epsilon < \frac{1}{16}$ There is an $O(nM(\Delta(G))$ NC^2 algorithm which, given a pair $A, B \subseteq E$, verifies that (A, B) is ϵ-Triad Regular or finds subsets $A' \subseteq A$ and $B' \subseteq B$ such that*

$$|\gamma(A', B') - \gamma(A, B)| > \frac{\epsilon^5}{90}$$

and $p(A', B') > \frac{\epsilon^5}{90} p(A, B)$.

Proof. We first compute $\gamma(A, B)$, which may be done in $O(n\Delta(G)^2)$ time and in NC. If $\gamma(A, B) < \epsilon$, $G(A, B)$ is trivially ϵ-Triad Regular and we are done.

Otherwise, we compute the set $Y \subseteq V_2$ of $(\frac{\epsilon^5}{90}, \frac{\epsilon^{5/2}}{9})$-triangle irregular vertices. This can be done in $O(n\Delta(G)^2)$ time by computing $|N(u) \cap N_B(v)|$ for each $v \in V_2$ and each $u \in N_A(v)$. We then check if $\text{pwt}^{A,B}(Y) > \frac{2\epsilon^{5/2}}{5} p(A, B)$, which may be done in time $O(n)$. If $\text{pwt}^{A,B}(Y) > \frac{2\epsilon^{5/2}}{5} p(A, B)$, we construct the following sets, each of which can be constructed in $O(n\Delta(G)^2)$ time.

For each $v \in Y$, let $A_v^+ \subseteq N_A(v)$ be the set $\{u \in N_A(v) : |N(u) \cap N_B(v)| - \gamma \deg_B(v) > \frac{\epsilon^5}{90} \deg_B(v)\}$ and let $A_v^- = \{u \in N_A(v) : |N(u) \cap N_B(v)| - \gamma \deg_B(v) < \frac{\epsilon^5}{90} \deg_B(v)\}$. Let $Y^+ = \{v \in Y : |A_v^+| \geq |A_v^-|\}$, and $Y^- = Y \setminus Y^+$. Assume without loss of generality that $\text{pwt}^{A,B}(Y^+) \geq \text{pwt}^{A,B}(Y^-)$. Define $A' = \bigsqcup_{v \in Y^+} \{vu : u \in A_v\}$ and $B' = B$. The parallelization of this construction is clear.

We now show that if $\mathrm{pwt}^{A,B}(Y) > \frac{2\epsilon^{5/2}}{5} p(A,B)$, then (A',B') is a $\frac{\epsilon^5}{90}$-Triad Irregular pair. We have

$$t(A',B') - \gamma p(A',B') = \sum_{v\in Y^+} \sum_{u\in A_v^+} |N(u) \cap N_B(v)| - \gamma \deg_B(v)$$

$$\geq \sum_{v\in Y^+} \sum_{u\in A_v^+} \frac{\epsilon^5}{90} \deg_B(v)$$

$$= \frac{\epsilon^5}{90} p(A',B')$$

and it follows that $\gamma(A',B') > \gamma(A,B) + \frac{\epsilon^5}{90}$. Similarly, we have

$$p(A',B') = \sum_{v\in Y^+} |A_v^+| \deg_B(v)$$

Since $v \in Y$, $|A_v^+| + |A_v^-| \geq \frac{\epsilon^{5/2}}{9} \deg_A(v)$. As A_v^+ is the larger of the two sets,

$$p(A',B') \geq \frac{\epsilon^{5/2}}{18} \sum_{v\in Y^+} \deg_A(v) \deg_B(v) = \frac{\epsilon^{5/2}}{18} \mathrm{pwt}^{A,B}(Y^+)$$

Since $\mathrm{pwt}^{A,B}(Y) \geq \frac{2\epsilon^{5/2}}{5} p(A,B)$ and $\mathrm{pwt}^{A,B}(Y^+) \geq \mathrm{pwt}^{A,B}(Y^-)$, we have

$$p(A',B') \geq \frac{\epsilon^{5/2}}{36} \sum_{v\in Y} \deg_A(v) \deg_B(v) = \frac{\epsilon^5}{90} p(A,B)$$

and thus (A',B') witnesses that (A,B) is $\frac{\epsilon^5}{90}$-Triad irregular.

Now suppose that both $\gamma(A,B) > \epsilon$ and $\mathrm{pwt}^{A,B}(Y) < \frac{2\epsilon^{5/2}}{5} p(A,B)$. Then we compute the local deviations $\sigma_v(A,B)$ for every vertex $v \in V_2$. This may be done in $O(n\Delta(G)^2)$ time and check if (A,B) has $(\frac{\epsilon^2}{9}, \frac{2\epsilon^2}{5})$-large deviations. This check can be done in $O(nM(\Delta(G)))$ time.

If (A,B) has $(\frac{\epsilon^2}{9}, \frac{2\epsilon^2}{5})$-large deviations, then by Lemma 2, there are subsets $A' \subseteq A$ and $B' \subseteq B$ such that $\gamma(A',B') - \gamma(A,B) > \frac{\epsilon^5}{90}$ and $p(A',B') > \frac{\epsilon^5}{90} p(A,B)$. Furthermore, these subsets may be found in $O(nM(\Delta(G)))$ time. Conversely, if (A,B) has $(\frac{\epsilon^2}{9}, \frac{2\epsilon^2}{5})$-small deviations, then Lemma 1 implies that (A,B) is ϵ-Triad Regular.

Each of the three steps may be performed in $O(nM(\Delta(G)))$ time, and thus the whole procedure requires $O(nM(\Delta(G)))$ time. $\square$

4   Algorithmic Clustering Regularity Lemma

For partitions $\mathcal{P}, \mathcal{Q}$ of E, we define the index function $I(\mathcal{P}, \mathcal{Q})$:

$$I(\mathcal{P},\mathcal{Q}) = \sum_{\substack{P_i\in\mathcal{P} \\ Q_j\in\mathcal{Q}}} \frac{p(P_i,Q_j)}{p(G)} \gamma(P_i,Q_j)^2$$

We will need the following standard facts regarding the index function, which are slight modifications of analogous results in [11–13].

Proposition 1. *The index function has the following properties:*

1. *For the partition $\mathcal{P}^{(0)} = (\{E\}, \{E\})$, $I(\mathcal{P}_0) = \gamma(G)^2$*
2. *If $\mathcal{P}'$ is a refinement of $\mathcal{P}$, then $I(\mathcal{P}') \geq I(\mathcal{P})$*
3. *$I(\mathcal{P}) \leq \frac{1}{2}$*

A partition $\mathcal{P}$ of E_{12}, E_{23} with parts $(\{P_1, \ldots, P_r\}, \{Q_1, \ldots, Q_s\})$ is ϵ-Triad Regular if

$$\sum_{\substack{P_i \in \mathcal{P} \\ Q_j \in \mathcal{Q} \\ (P_i, Q_j)\epsilon\text{-Triad irregular}}} p(P_i, Q_j) < \epsilon p(E_{12}, E_{23})$$

We will need one additional lemma which appears in the proof of [12, Theorem 4.3].

Lemma 3. *Let $(\mathcal{P}, \mathcal{Q})$ be partitions of E. Let $(\mathcal{P}', \mathcal{Q}')$ be a refinement of $(\mathcal{P}, \mathcal{Q})$ where if (P_i, Q_j) is an ϵ-Triad irregular pair $(\mathcal{P}, \mathcal{Q})$ with $A \subseteq P_i$ and $B \subseteq Q_j$ such that $|\gamma(A, B) - \gamma(P_i, Q_j)| > \delta$, then P_i is refined to $(A, P_i \setminus A)$ in $\mathcal{P}'$ and Q_j is refined to $(B, Q_j \setminus B)$ in $\mathcal{Q}'$. If $(\mathcal{P}, \mathcal{Q})$ is ϵ-irregular, then*

$$I(\mathcal{P}', \mathcal{Q}') \geq I(\mathcal{P}, \mathcal{Q}) + \epsilon^2 \delta^2$$

Finally, we can combine Lemmas 1 and 2 with the proof of the regularity lemma (Theorem 2) in [12] to construct our algorithm.

Theorem 3. *For every $\epsilon > 0$ there is a positive integer K such that any graph G with clustering coefficient γ and at least K edges has an ϵ-Triad Regular partition of of $G(E, E)$ into at most K parts. Furthermore, this partition may be constructed in NC^2 and in $O(nM(\Delta(G)))$ time, where $M(n)$ is the time needed to multiply two $n \times n$ matrices.*

Proof. Our algorithm proceeds as follows:

1. Construct the tripartite subcover graph $G(E, E)$, and set $\mathcal{P}_0 = (E, E)$
2. Given $(\mathcal{P}, \mathcal{Q})$ with parts $(P_1, \ldots, P_r, Q_1, \ldots, Q_r)$, apply the procedure from Corollary 1 to each pair (P_i, Q_j) to determine whether the pair is ϵ-Triad Regular or find a partition $P_{i1} \sqcup P_{i2} = P_i$ and $Q_{j1} \sqcup Q_{j2} = Q_j$.
3. If the total pathweight of the ϵ-Triad-irregular pairs is at most $\epsilon p(E, E)$, return the partition $\mathcal{P}_i$, otherwise refine the partition using the sets found in step 2 to form $\mathcal{P}_{i+1}, \mathcal{Q}_{i+1}$. Then return to step 2.

To show correctness, let $(\mathcal{P}, \mathcal{Q})$ be an ϵ-Triad irregular partition. Letting the partition computed in step 2 be $(\mathcal{P}', \mathcal{Q}')$, Lemma 3 and Corollary 1 imply that

$$I(\mathcal{P}', \mathcal{Q}') \geq I(\mathcal{P}, \mathcal{Q}) + \frac{\epsilon^{12}}{8100}.$$

Since $I(\mathcal{P}', \mathcal{Q}') \leq \frac{1}{2}$ by Proposition 1, the above algorithm terminates with an ϵ-Triad Regular partition after no more than $\frac{4050}{\epsilon^{12}}$ iterations.

Step 1 may be carried out in $O(n\Delta(G)^2)$ time. By Corollary 1, a given pair (P_i, Q_j) may be refined by step 2 in time $O(nM(\Delta(G)))$. As the size of the final partition is constant, there are a constant number of applications of Corollary 1, and thus step 2 as a whole requires $O(nM(\Delta(G)))$ time. Steps 3 and 4 may be performed in $O(n\Delta(G)^2)$ time. As the number of iterations is constant, the whole procedure requires $O(nM(\Delta(G)))$ time. $\qquad\qquad\square$

5 Further Directions

In applications of the regularity lemma, one often wishes to include the additional constraint that the partition is *equitable*, i.e. all parts have sizes as equal as possible. Our result can be extended to this setting, provided one defines two partitions to have equal size if they have equal *pathweight*. We leave the details to the interested reader.

While the algorithm runs in time polynomial is n, the leading constant has a tower-type dependence on ϵ, the same dependence which appears in Szemerédi's original lemma [13] and Theorem 2. As Alon and Moshkovitz show in [24], a tower-type dependence on ϵ is necessary in Chung's regularity lemma for clustering graphs, and we suspect an analogous more efficient result holds for Theorem 2. Freize and Kannan [25] give an alternative version of the regularity lemma with a weaker notion of regularity but vastly improved dependence on ϵ. It would be of interest to produce an analogous regularity lemma for clustering graphs.

While we cannot expect improved dependence on ϵ in general, the constructions in [24, 26] which give lower bounds are extremely delicate. It is entirely possible that graphs appearing in applications need not require such a vast number of parts in an ϵ-Triad Regular partition. We are keen to know then how many parts are actually needed in an ϵ-Triad Regular of a real-world clustering graph. Similarly we can for sufficient conditions which guarantee that a clustering graph has an ϵ-Triad Regular partition of size exponential in $\frac{1}{\epsilon}$. Both questions would be of interest.

References

1. Albert, R., Jeong, H., Barabási, A.L.: Diameter of the world-wide web. Nature **401**, 130–131 (1999)
2. Barabási, A.L., Albert, R., Jeong, H.: Scale-free characteristics of random networks: the topology of the world-wide web. XXPhys. A **281**, 69–77 (2000)
3. Watts, D.J., Strogatz, S.H.: Collective dynamics of 'small-world9 networks. Nature **393**, 440–442 (1998)
4. Steyvers, M., Tenenbaum, J.B.: The large-scale structure of semantic networks: statistical analyses and a model of semantic growth. Cogn. Sci. **29**, 41–78 (2005)
5. Faust, T.Y., Porter, M.A.: Inference of hierarchical core-periphery structure in temporal network (2025). arXiv:2506.10135 [cs]

6. Leskovec, J., Lang, K.J., Dasgupta, A., Mahoney, M.W.: Statistical properties of community structure in large social and information networks. In: Proceedings of the 17th International Conference on World Wide Web, WWW 2008, (New York, NY, USA), pp. 695–704. ACM (2008)

7. Bourla, G., Wang, K., Wei, F., Zhou, R.: Social networks: enumerating maximal community patterns in c-Closed Graphs (2025). arXiv:2506.11437 [math]

8. Simas, T., Rocha, L.M.: Distance closures on complex networks. Netw. Sci. **3**, 227–268 (2015)

9. Dougherty, D.P., Stahlberg, E.A., Sadee, W.: Network analysis using transitive closure: new methods for exploring networks. J. Stat. Comput. Simul. **76**, 539–551 (2006). _eprint: https://doi.org/10.1080/10629360500107857

10. Gupta, R., Roughgarden, T., Seshadhri, C.: Decompositions of triangle-dense graphs. SIAM J. Comput. **45**, 197–215 (2016)

11. Chung, F.: Regularity lemmas for clustering graphs. Adv. Appl. Math. **126**, (2021)

12. Chung, F., Sieger, N.: Eigenvalues of clustering graphs. J. Linear Algebra Appl. (2025)

13. Szemerédi, E.: On sets of integers containing no k elements in arithmetic progression. Acta Arith **27**, 199–245 (1975)

14. Gowers, W.T.: Hypergraph regularity and the multidimensional Szemerédi theorem. Ann. Math. **166**(3), 897–946 (2007). arXiv: 0710.3032

15. Vadhan, S.P.: Pseudorandomness. Found. Trends Theor. Comput. Sci. **7**(1–3), 1–336 (2011)

16. Komlos, J., Simonovits, M.: Szemeredi's regularity lemma and its applications in graph theory, technical report, Center for Discrete Mathematics & Theoretical Computer Science (1995)

17. Nagle, B., Rödl, V., Schacht, M.: The counting lemma for regular k-uniform hypergraphs. Random Struct. Algorithms **28**, 113–179 (2006)

18. Alon, N., Duke, R.A., Lefmann, H., Rodl, V., Yuster, R.: The algorithmic aspects of the regularity lemma. J. Algorithms **16**, 80–109 (1994)

19. Lubotzky, A.: High dimensional expanders (2017). arXiv:1712.02526

20. Oppenheim, I.: Local spectral expansion approach to high dimensional expanders part i: descent of spectral gaps. Discrete Comput. Geom. **59**, 293–330 (2018)

21. Oppenheim, I.: Local spectral expansion approach to high dimensional expanders part ii: mixing and geometrical overlapping. Discrete Comput. Geom. **64**, 1023–1066 (2020)

22. Chung, F.R., Graham, R.L., Wilson, R.M.: Quasi-random graphs. Combinatorica **9**, 345–362 (1989)

23. Chung, F., Sieger, N.: Quasi-random clustering graphs (2026), in preparation

24. Alon, N., Moshkovitz, G.: Limitations on regularity lemmas for clustering graphs. Adv. Appl. Math. **124**, 102135 (2021)

25. Frieze, A., Kannan, R.: A simple algorithm for constructing Szemerédi's regularity partition. Electron. J. Comb. R17–R17 (1999)

26. Gowers, W.: Lower bounds of tower type for Szemerédi's uniformity lemma. Geom. Funct. Anal. GAFA **7**, 322–337 (1997)

Fast, Parallel, and Scalable Differential Graph Compression

Davide Cologni[1] and Sebastiano Vigna[2]($\boxtimes$)

[1] Dipartimento di Scienze Ambientali, Informatica e Statistica,
Università Ca' Foscari Venezia, Venice, Italy
`davide.cologni@unive.it`
[2] Dipartimento di Informatica, Università Degli Studi di Milano, Milan, Italy
`sebastiano.vigna@unimi.it`

Abstract. Graph compression is essential for storing and analyzing large-scale graphs such as web crawls, social networks, and code repositories. In this paper we discuss new compression techniques for the Rust implementation of the WebGraph framework [9]. We take inspiration from Zuckerli's [13] extensions to the classical Boldi–Vigna (BV) differential graph compression scheme, and propose a scalable, parallelizable compressor. We also develop a hybrid Huffman coding implementation with programmable maximum codeword length which makes it possible to eliminate context modeling and achieve faster decoding times with similar compression effectiveness. We obtain compression ratios up to 15% better than the standard BV compressor, while providing random-access times up to 54% faster than the Zuckerli implementation. We combine these techniques with a new, streamlined version of Apostolico–Drovandi's π codes, which further improve compression and increase encoding and decoding speed. We validate our approach on graphs ranging from hundreds of thousands to over a billion nodes, including a Software Heritage archive with ≈ 50 billion nodes and ≈ 854 billion arcs. In particular, we show that, unlike Zuckerli, we can improve at the same time both the compression ratio and the random-access speed with respect to WebGraph.

Keywords: graphs · compression · data structures · prefix-free codes

1 Introduction

Large real-world data instances such as social networks, web snapshots, and code repositories can be naturally represented as graphs. Due to the increasing size of

This work is partially supported by the project "SEcurity and RIghts In the CyberSpace - SERICS" (PE00000014 - CUP H73C2200089001) under the National Recovery and Resilience Plan (NRRP) funded by the European Union - NextGenerationEU. Views and opinions expressed are those of the authors only and do not necessarily reflect those of the European Union or the Italian MUR. Neither the European Union nor the Italian MUR can be held responsible for them.

F. C. Graham et al. (Eds.): WAW 2026, LNCS 16630, pp. 97–111, 2026.
https://doi.org/10.1007/978-3-032-27193-8_7

such graphs, compressed graph representations have become essential for analysis and distribution.

WebGraph [5] is a compression framework developed in 2004 that exploits the structural redundancy of web graphs: more specifically, the tendency of pages to link to pages with a nearby index (locality) and for related pages to share successors (similarity). Originally implemented in Java and recently ported to Rust [9], WebGraph has been widely adopted for web crawling by Common Crawl, social network analysis, and large-scale graph processing applications [4].

Zuckerli [13] improves upon WebGraph by introducing statistical coding with context modeling and a dynamic programming approach to reference selection. These techniques achieve 10–29% better compression ratios compared to WebGraph's instantaneous codes. However, the reference implementation has significant limitations: its global reference selection algorithm is memory-hungry and inherently single-threaded, and its contextualized hybrid Huffman encoder/decoder is quite slow.

Our main motivation comes from graph management at Software Heritage (SWH) [8], a UNESCO-sponsored project aiming at storing all existing software in a DAG that keeps track of change history. The current snapshots of SWH have almost a trillion arcs, and while SWH has been using WebGraph for a long time, improving compression techniques in a scalable way has an immediate positive impact.

The main contributions of this paper are a new parallel reference selection algorithm; a new hybrid Huffman encoder with configurable codeword length limits; and an alternative to Apostolico–Drovandi π codes [1] (a family of universal codes) that improves encoding and decoding speed.

Our implementation achieves compression ratios close to Zuckerli (within 2% on most graphs) while providing 25–54% faster random-access times and enabling parallel compression. More importantly, we show that we can provide both better compression ratios and faster random-access times than WebGraph, thus overcoming the fundamental trade-off observed in the original Zuckerli paper.

The code for our new compressor has been integrated into the Rust port of WebGraph;[1] we also make available our new Huffman compressor.[2] Streamlined π codes are implemented in the Rust crate `dsi-bitstream`.[3] All the code is open source under either the Apache 2.0 or LGPLv2 license.

2   WebGraph

The WebGraph framework is the standard benchmark for graph compression. For a survey of related work and alternative approaches, we refer to the up-to-date related-work section of the Zuckerli paper [13].

WebGraph relies on two common properties of real-world graphs: *locality* and *similarity*. Locality means that the successors of a node often have IDs that are

---

[1] https://github.com/vigna/webgraph-rs.
[2] https://github.com/colobrodo/webgraph-rs.
[3] https://crates.io/crates/dsi-bitstream.

numerically close to each other. Similarity indicates that nodes with similar IDs often share many successors.

To exploit similarity, WebGraph uses *differential compression*. Instead of storing the full successor list for a node u, denoted as $S(u)$, the framework encodes it relative to a *reference node r*. This reference is chosen from a sliding window of previously processed nodes, $\{u - 1, \ldots, u - W\}$. The set of common successors $S(u) \cap S(r)$ is stored using *copy blocks*: a sequence of lengths that indicate which contiguous parts of r's list should be copied or skipped.

The successors of u that are not found in r are called *residuals*. To further exploit locality, the framework searches in the residuals for sufficiently long *intervals*. These intervals are encoded using their starting value and length. Any remaining residuals are stored using *gap encoding* (i.e., storing the difference between consecutive values). The first residual is encoded as the difference relative to the node u itself.

WebGraph selects the reference node r using a greedy strategy. For each node u, it tests all candidates in the window and chooses the one that results in the smallest representation of $S(u)$. This approach creates *reference chains*: to decode node u, one may need to decode r, which might depend on another node, and so on. Although long chains improve compression, they slow down random access. WebGraph limits the maximum length of these chains to a fixed parameter R.

3   Zuckerli

To provide the necessary context for the proposed improvements, we detail some of the inner workings of the Zuckerli compression scheme. The structural decomposition of successor lists remains unchanged from WebGraph, with the exception of a slightly different treatment of intervals. However, Zuckerli uses a new hybrid Huffman encoding and a larger number of codes (e.g., it uses a different code for the first residual and for the remaining residuals). More importantly for us, it introduces a better algorithm for reference selection.

3.1   Hybrid Integer Encoding

A major challenge to the application of statistical encoding in graph compression is the cardinality of the symbol alphabet. For a graph with n nodes, a naive encoding of successor lists implies an alphabet size of n, making probability tables too large to store and construct efficiently.

The encoding exploits locality, as arcs tend to connect nodes with similar indices. As a result, the generated integers follow a distribution heavily skewed towards small values. To make statistical encoding feasible, Zuckerli uses *hybrid integer encoding*, parameterized by i, j, and k (with $i \geq j + k$). This technique reduces the alphabet size by applying the statistical model to only a portion of the integer binary representation.

For an integer x, the encoding proceeds as follows:

1. If $x < 2^i$, the value is encoded directly as a symbol $s = x$. This preserves the precise statistical properties of frequent, low-magnitude values.
2. If $x \geq 2^i$, the integer is decomposed. The encoding excludes the most significant bit, and then extracts the j least significant bits and the k most significant bits of the remaining value. The middle segment, comprising t bits, is emitted verbatim to the bitstream. The statistical symbol generated for x encodes the triplet comprising the j LSBs, the k MSBs, and the length t.

This reduces the alphabet size by excluding the middle bits from the model.

3.2   Context Modeling

To enhance compression effectiveness, Zuckerli uses context modeling to select probability distributions based on previously processed symbols, exploiting their statistical correlations.

For node outdegrees, the context is derived from the outdegrees of previous nodes. To mitigate sequential dependencies that would impede random access, outdegrees are partitioned into fixed-size blocks (e.g., 32 elements). The first element of each block uses a default context, which effectively severs the dependency chain.

Reference encoding employs the same modeling strategy. The context for a block depends on whether it is the first block, an even-indexed block, or an odd-indexed block. Finally, residual encoding adapts to the position within the list. The first residual depends on the total residual count, and subsequent ones depend on their immediate predecessor.

3.3   Approximate Reference Selection

The effectiveness of reference encoding relies on selecting an optimal reference r for each node u. The greedy heuristic used by WebGraph–which selects the candidate minimizing the resulting encoded bit-length–is efficient to compute but fails to capture the utility of u as a reference for future nodes.

Zuckerli improves upon this by proposing a $\left(1 - \frac{1}{R+1}\right)$-approximation algorithm, in which reference selection is modeled as the problem of finding the maximum R-bounded forest in a weighted DAG. Here, an arc (r, u) exists if and only if r is a valid reference for u (i.e., r is among the W predecessors of u). The weight of an arc, $w(r, u)$, represents the compression gain achieved by encoding u relative to r versus encoding u autonomously. In this context, the parameter R imposes a constraint on the maximum depth of the reference trees.

The approximation algorithm proceeds in three phases:

1. A maximum-weight forest F is first constructed by selecting, for each node u, the reference r that maximizes $w(r, u)$, disregarding the depth constraint.
2. Then a dynamic programming procedure is applied to extract a sub-forest $F' \subseteq F$ of maximum weight such that no path in F' exceeds length R.

3. Finally the forest F' is augmented by greedily inserting valid arcs (r, u) that strictly increase the total weight without violating the depth constraint R, thereby recovering a portion of the compression gain discarded during pruning.

Note that in the first step the reference maximizing $w(r, u)$ can always be chosen greedily for each node, since there is no depth constraint. Indeed, this step is effectively equivalent to the WebGraph greedy algorithm with $R = +\infty$, because without depth constraints the global optimum decomposes into independent local choices.

A subtle circular dependency arises between the reference selection and the statistical encoder. The weight $w(r, u)$ relies on the bit-length of the encoded residuals, which in turn depends on the symbol frequencies resulting from the reference selection. Zuckerli resolves this through an iterative process using a sequence of cost models. The initial iteration assumes a uniform bit-cost for all symbols, while subsequent iterations refine the cost function based on the symbol frequencies observed in the previous step.

4   Fast, Parallel, and Scalable Differential Compression

We now describe in more detail our new scalable compressor framework, highlighting the differences from Zuckerli.

4.1   Hybrid Huffman Encoder

We provide a new hybrid Huffman encoder with a configurable maximum codeword size. Zuckerli hardcodes a maximum codeword length of 8 bits, using three bits to store lengths in the header. A programmable maximum codeword size enables the creation of decoding tables that speed up decoding times, albeit at the expense of the compression ratio. Our implementation supports arbitrary codeword length limits, provided that the underlying bitstream allows for peeking enough bits. In our experiments we use 12 bits, which provides better compression while maintaining fast, table-based decoding.

We also provide two context model implementations: a *constant* model (single context) and a *per-component* model (separate distributions for outdegree, reference, blocks, intervals, and residuals). Both models achieve slightly worse compression ratios than the model proposed by Zuckerli but can achieve faster decoding times thanks to two main factors:

- the decoding table constructed is significantly smaller, reducing the number of cache misses during decoding;
- a fixed distribution for outdegrees and references allows us to avoid the overhead of context resolution, that is, decoding the outdegrees and references of previous lists in the block.

Simpler models paired with longer Huffman words yield a decoder that maintains the same performance as the current WebGraph implementation but reduces the size of the compression output.

4.2 Chunked Reference Selection

The original Zuckerli reference selection algorithm operates on the global graph structure, maintaining a dynamic programming table that scales with $O(nR)$. This approach imposes a prohibitive memory footprint for massive graphs and creates a sequential dependency that precludes parallelization.

To address these scalability constraints, we adopt a chunk-based strategy. We divide the graph into fixed-size *chunks* (sequences of consecutive nodes) and execute the approximate reference selection algorithm independently within each chunk. This architectural adaptation allows us to control the memory usage of the compressor and enables several optimizations.

First, within each chunk, we employ a ragged array (a flat array storing variable-length rows contiguously) to store the successor lists in the chunk. This compact representation significantly improves memory locality compared to standard vector-of-vectors approaches, allowing the algorithm to scale efficiently even with larger chunk sizes.

Second, restricting reference selection to a single chunk enables us to cache the encoding cost of each list with each possible reference. Unlike the Zuckerli implementation, which re-evaluates costs during the greedy augmentation phase, our approach stores these computed weights, thereby eliminating redundant computations during the final greedy phase.

Most importantly, because chunks are independent, the compression process is trivial to parallelize. This enables our implementation to effectively utilize multi-core processors and scale to graphs containing hundreds of billions of arcs, which the single-threaded global approach cannot handle in terms of both memory and compression time.

4.3 Iterative Estimation

As discussed in Sect. 3.3, the combination of Huffman coding with reference selection results in a circular dependency: the Huffman encoder needs to know in advance the frequencies of each symbol to determine its codewords, and the reference selection algorithm needs to know the bit size of the encoded lists to determine the most effective reference.

We have also improved the iterative process that Zuckerli uses to resolve this dependency. More precisely, the first round employed by Zuckerli assumes a uniform bit-cost for all symbols; however, it is well established that gaps in successor lists follow Zipfian-like distributions and small symbols occur significantly more often than others. This knowledge is also exploited by the implementation of hybrid integer encoding, which assigns unique symbols to smaller integers.

We choose a better estimator for the first round that results in a cost of $\lfloor \lg(x + 2) \rfloor$ (where lg denotes the base-2 logarithm) for the integer x to encode. Another distinction between our estimation procedure and the reference implementation is that we recompute references in the final round using the last obtained estimator, whereas Zuckerli reuses the references computed in the penultimate round and applies the final cost model only during encoding; thus, we perform $t+1$ reference computations for t estimation rounds versus Zuckerli's t. This final round is necessary because our chunked approach discards partial results as the graph is being compressed, but this increase in computation time is largely offset by our parallel approach. (An alternative approach would involve persisting to disk the previously computed references.)

5 Streamlined π Codes

Apostolico and Drovandi [1] devised π codes in the context of breadth-first search graph compression. Similarly to ζ codes [6], these codes were designed to exploit the specific power-law distribution of gaps between neighboring nodes often found in web graphs, as the π code with parameter $k \geq 0$ for a natural number n has an implied probability distribution $\approx 1/n^{1+1/2^k}$. A ζ code with parameter $k \geq 1$ has instead a distribution $\approx 1/n^{1+1/k}$, which makes it possible to model distributions more accurately: at the same time, ζ codes are more onerous than π codes to encode and decode, as they need to handle minimal binary codes.

In the original paper, the definition of the code is algorithmically complex, involving a construction that obscures its relationship with other well-known universal codes. We thus describe an alternative set of codes, termed *streamlined π codes*, whose codewords are different from those of π codes, but whose codeword for each n has the same length.

Specifically, the *streamlined π code* with parameter $k \geq 0$ for a natural number n consists of the juxtaposition of two parts:

1. the Golomb code [10] with parameter 2^k of $\lfloor \lg(n+1) \rfloor$ (equivalently, a Rice code [12] with parameter k);
2. the binary representation of $n + 1$ with the most significant bit removed.

Note that replacing the first part with a unary code we obtain the Elias γ code, and replacing it with a γ code we obtain the Elias δ code.

Table 1 illustrates this construction. While the codeword lengths produced by this definition are identical to those in the original paper (ensuring they remain statistically equivalent), the actual bit patterns for $k \geq 1$ are different. However, this structural simplification allows for faster encoding and decoding routines, as it leverages standard primitives for Rice coding and binary representations, reducing the number of instructions required. More precisely, in the original definition the initial unary prefix of length u is interpreted as the integer u (rather than $u - 1$, as we do), and then the following k bits do not represent a residual, as in a Golomb code, but a value to be subtracted from $u \cdot 2^k$.

While technically this improvement is orthogonal to our new compression engine, the two interact synergistically, as we will show in the next section.

6 Experimental Evaluation

We evaluate compression ratio, compression time, and random-access speed. All experiments use `rustc` 1.92.0 and `clang` 19.1.7 on Linux and are performed on

Table 1. Construction of streamlined π codes with parameter $k = 1, 2$. The first part of the code for n is formed by the Golomb code with parameter 2^k of $\lfloor \lg(n + 1) \rfloor$, concatenated with the binary representation of $n + 1$ with the most significant bit removed.

n	$n + 1$	$\lfloor \lg(n+1) \rfloor$	Streamlined π_1	Original π_1	Streamlined π_2	Original π_2
0	1_2	0	10|	11|	100|	111|
1	10_2	1	11|0	10|0	101|0	110|0
2	11_2	1	11|1	10|1	101|1	110|1
3	100_2	2	010|00	011|00	110|00	101|00
4	101_2	2	010|01	011|01	110|01	101|01
5	110_2	2	010|10	011|10	110|10	101|10
6	111_2	2	010|11	011|11	110|11	101|11
7	1000_2	3	011|000	010|000	111|000	100|000

an Intel(R) Core(TM) i7-12700KF @ 3.60 GHz machine with 64 GB of RAM and 20 cores. We execute our experiments on graphs from the Laboratory for Web Algorithmics[4] spanning different domains (Table 2):

- uk-2005: a 2005 crawl of the .uk domain;
- dblp-2011: an undirected scientific collaboration network;
- imdb-2021: a graph of movie actors; vertices are actors, and two actors are joined by an edge if they appeared in a movie together;
- enwiki-2025: a snapshot of the English part of Wikipedia collected in 2025 [2];
- twitter-2010: a crawl presented in [11];
- ljournal-2008: the snapshot of LiveJournal used in [7];
- eu-2015: a large snapshot of pages from European domains gathered by BUb-iNG [3];
- swh-2025: the 2025-10-08 snapshot of the Software Heritage graph, representing the full DAG of the world's largest source code archive.

6.1   Chunked Reference Selection Compression

We empirically confirm that the Zuckerli reference selection algorithm only incurs a negligible decrease in compression ratio when executed in chunks. For instance, on enwiki-2025, changing the chunk size from 10 to 1 000 000 nodes alters the compression ratio by less than 3%, with differences below 0.003% for chunks with more than 10 000 elements.

6.2   Starting Estimation Model

As part of our evaluation process, we measure the accuracy of our initial estimator on several graphs. We then compare the resulting compressed size after

---

[4] https://law.di.unimi.it/datasets.php.

Table 2. Graphs used during the experimental evaluation.

Graph	Nodes	Arcs
uk-2005	39 459 925	936 364 282
dblp-2011	986 324	6 707 236
imdb-2021	2 996 317	10 739 291
enwiki-2025	6 961 383	181 022 291
twitter-2010	41 652 230	1 468 365 182
ljournal-2008	5 363 260	79 023 142
eu-2015	1 070 557 254	91 792 261 600
swh-2025	49 903 891 086	853 845 713 989

Table 3. Graph size (measured in bits per arc) obtained with different rounds and different starting cost models.

Graph	rounds	1		2		3	
	starting estimator	fixed	log	fixed	log	fixed	log
uk-2005		1.603	1.545	1.539	1.538	1.541	1.541
dblp-2011		8.335	7.598	7.552	7.543	7.539	7.542
imdb-2021		8.751	8.236	8.185	8.169	8.169	8.169
ljournal-2008		10.544	10.106	10.072	10.071	10.071	10.071
twitter-2010		13.233	12.802	12.773	12.773	12.772	12.773
enwiki-2025		12.628	12.023	11.997	11.996	11.996	11.996

different rounds, starting either with the fixed cost model used in the Zuckerli reference implementation or with our alternative model, which is based on the binary representation of the integer. The results in Table 3 show that, although the two configurations achieve very similar compression ratios in the second round, our alternative achieves a result much closer to the final outcome in the first round. This makes single-round compression a viable option for faster compression, particularly for larger graphs where encoding times become prohibitive.

6.3   Context vs. Codeword Length Trade–Off

Context modeling allows us to capture and exploit statistical correlations in the data. However, increasing the number of contexts also requires storing additional distributions and, most importantly, creating larger decoding tables. For a given codeword limit of l, the decode table for each context contains exactly 2^l entries. Therefore, each additional codeword bit doubles the overall decode table size. Alternatively, one can double the number of contexts while reducing the codeword length by one bit and keeping the decode table occupancy constant. We investigated this trade-off using a set of integers taken from the residuals of different graphs. We measured both the resulting compressed file size and the

Table 4. The resulting file sizes in bytes for residual compression with different contexts and codeword limits.

	1 ctx 12 bits	2 ctx 11 bits	4 ctx 10 bits	8 ctx 9 bits
uk-2007-02	2464M	2462M	**2453M**	2478M
enwiki-2023	1757M	1754M	1751M	**1748M**
eu-2005	35.1M	35.0M	34.9M	**34.8M**

sequential decoding time. The selection of contexts for residuals in this experiment is performed similarly to what Zuckerli did: the first residual is assigned to a fixed context, while the contexts for successive residuals, encoded as deltas, are determined by the previous symbols used.

Generally, more contexts improve compression (Table 4) but increase decoding time: with 12 bits and 1 context the sequential read time is 7.4–7.7 ns/symbol. On the opposite end of the spectrum, 9 bits with 8 contexts require 7.6–8.2 ns/symbol due to poor cache locality when switching between tables. We opted for 12-bit codewords with per-component contexts, sacrificing some compression for faster decoding times.

6.4   Experimental Setup

To estimate the effectiveness of our compression techniques, we compare the Zuckerli reference implementation with several configurations of our design.

More precisely, we compare the standard WebGraph compressor with instantaneous codes and both the greedy and chunked reference selection algorithms (with a chunk size of 10 000). For convenience, these are named bv and bvz, respectively. We then compare the respective versions of these reference selection algorithms with our Huffman encoder, called huff and bvz-huff. Finally, we consider two configurations of WebGraph that use our streamlined π_2 codes to encode the residuals instead of the standard ζ_3 codes [6] with both reference selection strategies: pi2 and bvz-pi2.

We use the default window parameter $W = 7$ for the WebGraph greedy reference selection. For Zuckerli and our improved compressor, we change the window parameter to $W = 16$ for the web graphs (uk-2005 and eu-2015), for Twitter (twitter-2010), and for Software Heritage (swh-2025); we discuss the reasons for this choice in Sect. 6.8.

We leave the default minimum interval length $L = 4$ and the maximum reference chain length $R = 3$ for all the compressors. The Huffman encoders we use, including our own, adopt the Zuckerli default hybrid integer encoding parameters $i = 1, j = 2, k = 4$. Additionally, we deviate from the 8-bit hardcoded limit of the Zuckerli compressor by setting the maximum codeword length to 12 bits.

In all our tables, missing entries for Zuckerli indicate that the compressor ran out of memory.

Table 5. Compression ratio measured in bits per arc (relative improvement over bv in parentheses).

Graph	bv	Compressors					
		bvz	pi2	bvz-pi2	huff	bvz-huff	zuckerli
dblp-2011	8.71	8.63 (−1%)	8.53 (−2.1%)	8.45 (−3.1%)	7.77 (−10.8%)	7.66 (−12.1%)	7.59 (−12.8%)
imdb-2021	8.76	8.74 (−0.4%)	8.61 (−1.7%)	8.59 (−2%)	8.24 (−5.9%)	8.21 (−6.4%)	7.99 (−8.7%)
ljournal-2008	11.07	11.02 (−0.6%)	10.75 (−2.9%)	10.70 (−3.5%)	10.26 (−7.3%)	10.20 (−8%)	10.07 (−9.0%)
enwiki-2025	13.42	13.29 (−1.1%)	13.00 (−3.1%)	12.87 (−4.2%)	12.34 (−8.1%)	12.19 (−9.3%)	12.07 (−10.0%)
uk-2005	1.79	1.74 (−2.8%)	1.76 (−1.7%)	1.72 (−4.2%)	1.61 (−10.1%)	1.56 (−12.9%)	1.48 (−17.4%)
twitter-2010	14.11	13.95 (−1.1%)	13.64 (−3.3%)	13.49 (−4.4%)	13.02 (−7.7%)	12.83 (−9.1%)	12.64 (−10.4%)
eu-2015	1.19	1.13 (−5.1%)	1.17 (−1.6%)	1.11 (−6.9%)	1.07 (−10.4%)	1.01 (−15.2%)	—
swh-2025	2.00	1.98 (−1.0%)	1.95 (−2.5%)	1.94 (−3.0%)	1.82 (−9%)	1.80 (−9.89%)	—

6.5 Compression Ratio

As stated in the introduction, our goal is to avoid a trade-off between compression ratio and decoding speed. Indeed, Table 5 shows that in all cases our improved compressor provides better compression than WebGraph, and in combination with our Huffman encoder approaches Zuckerli's compression ratios. In all cases, streamlined π_2 codes improve over the default ζ_3 codes currently used by WebGraph.

The remaining gap versus Zuckerli stems partly from its complex context modeling, but also from a number of specific local improvements (such as a different code for the first residual). While it would be trivial to add these modifications, they would make the compressed graph backward-incompatible with WebGraph, and in our main use cases compatibility is a key requirement.

6.6 Compression Time

Table 6. Compression time in seconds. For each graph, the first row shows sequential time and the second row shows parallel time using all 20 available cores. `swh-2025` has been compressed only in parallel.

Graph	bv	bvz	huff	bvz-huff	zuckerli
			Compressors		
dblp-2011	1.12	1.39	3.12	3.83	3.96
	0.11	0.12	0.31	0.35	–
imdb-2021	1.83	2.14	5.47	6.25	8.32
	0.31	0.49	0.97	1.05	–
ljournal-2008	10.33	13.15	30.78	36.71	43.44
	1.50	2.49	4.98	5.79	–
enwiki-2025	22.20	31.07	61.17	85.84	88.79
	2.37	4.61	8.17	11.36	–
uk-2005	28.64	88.55	88.73	256.20	316.15
	3.89	8.79	10.07	24.18	–
twitter-2010	180.36	419.55	477.25	1164.08	1668.05
	63.21	150.31	175.97	411.93	–
eu-2015	2121.63	6065.68	5577.28	17002.08	–
	333.34	511.88	492.61	1324.49	–
swh-2025	7701.00	8436.77	13466.11	23215.97	–

We evaluate the scalability of the compression process by comparing the execution time of our chosen configurations on a single thread with Zuckerli's sequential execution (Table 6). Since our implementation can be parallelized by processing chunks independently, while Zuckerli's global reference selection is inherently sequential, we also measure the performance of compression on all available cores.

When running single-threaded, our implementation is only 3–30% faster than Zuckerli on larger graphs. This is mainly due to the fact that our chunked implementation requires an additional round of estimation (three instead of Zuckerli's two), which means that the code itself is actually almost twice as fast. Once we consider parallel execution, our implementation becomes several times faster, which was our main goal, and makes it possible to easily compress the large graphs such as `eu-2015` and `swh-2025`, for which Zuckerli runs out of memory. As expected, the new compressor is slower than WebGraph's greedy compressor; the gap is particularly wide for the graphs with $W = 16$ as the WebGraph compressor is always run with $W = 7$.

6.7   Random–Access Speed

Finally, we measure the speed of enumeration of successors. We pick $1\,000\,000$ lists uniformly at random, decode them, and measure the average time per arc. We report the minimum over 20 repetitions of the experiment, ensuring enough warm-up rounds.

Table 7. Random-access time in ns/arc. The first percentage in parentheses refers to bv, while the second value in the `bvz-huff` column compares to `zuckerli`.

| Graph | bv | Compressors | | | | | |
		bvz	pi2	bvz-pi2	huff	bvz-huff	zuckerli
dblp-2011	34.29	31.53 (-8%)	32.10 (-6%)	<u>30.05</u> (-12%)	72.65 (+112%)	67.08 (+96%, -47%)	128.76 (+276%)
imdb-2021	34.74	33.65 (-3%)	32.84 (-5%)	<u>31.59</u> (-9%)	68.47 (+97%)	66.61 (+92%, -53%)	143.36 (+313%)
ljournal-2008	21.40	20.41 (-5%)	20.15 (-6%)	<u>19.08</u> (-11%)	52.13 (+144%)	49.37 (+131%, -45%)	89.13 (+316%)
enwiki-2025	22.52	19.71 (-12%)	20.04 (-11%)	<u>17.74</u> (-21%)	54.78 (+143%)	50.73 (+125%, -42%)	87.39 (+288%)
uk-2005	21.36	18.32 (-14%)	21.25 (-1%)	<u>18.00</u> (-16%)	45.23 (+112%)	37.70 (+76%, -33%)	56.93 (+167%)
twitter-2010	15.22	14.55 (-4%)	14.02 (-8%)	<u>13.64</u> (-10%)	45.38 (+198%)	44.85 (+195%, -25%)	60.04 (+294%)
eu-2015	9.10	7.83 (-14%)	8.76 (-4%)	<u>7.58</u> (-17%)	27.56 (+203%)	23.10 (+154%, ——)	—
swh-2025	7.42	7.09 (-4%)	7.11 (-4%)	<u>6.77</u> (-9%)	23.39 (+215%)	23.35 (+214%, ——)	—

As shown in Table 7, graphs compressed using `bvz-huff` are 25–53% faster than Zuckerli.

More importantly, `bvz`, our compressor with instantaneous codes and chunked reference selection, outperforms WebGraph's default version on all graphs. This happens because even though the algorithm aims to minimize the bit-size representation of a list during the selection of a reference, it often does so by choosing an encoding that is more efficient in terms of total written integers, which reduces the number of operations needed by the decoder to reconstruct the original list. Finally, our streamlined π codes consistently make decoding faster. In particular, the combination `bvz-pi2` consistently achieves the best random-access speed across all graphs, while also improving compression by 2–7% over `bv`. We thus have a new compressor delivering both better compression and faster random access, at the price of slower compression.

These tests are somewhat favorable to Zuckerli, which uses an eager approach to list decoding: it first reconstructs the entire list in memory, and then serves requests from there; WebGraph uses a lazy approach that avoids allocating memory proportional to the list size, which is essential when manipulating large graphs, but is slightly slower.

6.8   Choosing the Right Window Size

Throughout our experiments, we observed an important pattern that has gone unnoticed both in WebGraph and in Zuckerli: large window sizes can have a negative impact on cache behavior. Indeed, large window sizes allow the reference selection algorithm to find better matches, but at the same time copying from distant references can generate many cache misses during decompression. The effect is much less pronounced in highly compressible graphs of low degree (such as web graphs) because the lists themselves end up being very short. Shortening the window is not always a solution; for example, in the case of the Software Heritage graph the Zuckerli reference-selection algorithm generates a slightly larger compressed representation ($+5\%$) when run with the default $W = 7$. While this result might seem surprising, one must remember that the algorithm first finds an optimal assignment for the case of infinite-length reference chains, and then prunes it: sometimes this process generates a choice that is worse than the greedy one. Our choice of window sizes tries to balance these two effects.

7   Conclusions

We propose a new compressor for graphs based on the WebGraph framework and ideas from Zuckerli that consistently outperforms WebGraph both in compression ratio and in random-access speed, at the price of a longer compression time, which however can be mitigated with parallelization: the Software Heritage graph we consider, which has ≈ 854 billion, can be compressed with our chunked reference selection in about two and a half hours. In the same vein, streamlined π_2 codes offer better compression and decoding speed with respect to the default WebGraph choice.

We carefully introduced these new ideas preserving backward compatibility and considering parallelization and memory footprint: these considerations are essential for real-world applications, such as those at Software Heritage.

In our experiments, we highlighted a previously overlooked interaction between the compression window size and random-access speed: a larger window improves compression and, by copying more arcs from predecessors, speeds up decompression; however, it also increases cache misses, making decompression slower.

In the future, we plan to perform a more thorough analysis of the window-sizing problem to balance these two forces, and consider alternatives to our current three-round evaluation of references in the Huffman case (e.g., by offline caching).

References

1. Apostolico, A., Drovandi, G.: Graph compression by BFS. Algorithms **2**(3), 1031–1044 (2009)
2. Boldi, P., Furia, F., Vigna, S.: Ten years of open Wikipedia ranking. In: Companion Proceedings of the ACM on Web Conference 2025, WWW 2025, pp. 883–887. ACM
3. Boldi, P., Marino, A., Santini, M., Vigna, S.: BUbiNG: massive crawling for the masses. ACM Trans. Web **12**(2), 12:1–12:26 (2019)
4. Boldi, P., Pietri, A., Vigna, S., Zacchiroli, S.: Ultra-large-scale repository analysis via graph compression. In: 2020 IEEE 27th International Conference on Software Analysis, Evolution and Reengineering (SANER), pp. 184–194. IEEE (2020)
5. Boldi, P., Vigna, S.: The WebGraph framework I: Compression techniques. In: Proceedings of the Thirteenth International World Wide Web Conference (WWW 2004), pp. 595–601. ACM Press, Manhattan, USA (2004)
6. Boldi, P., Vigna, S.: Codes for the world-wide web. Internet Math. **2**(4), 405–427 (2005)
7. Chierichetti, F., Kumar, R., Lattanzi, S., Mitzenmacher, M., Panconesi, A., Raghavan, P.: On compressing social networks. In: KDD 2009: Proceedings of the 15th ACM SIGKDD, pp. 219–228. ACM, New York, NY, USA (2009)
8. Di Cosmo, R., Zacchiroli, S.: Software Heritage: why and how to preserve software source code. In: Proceedings of the 14th International Conference on Digital Preservation, iPRES 2017 (2017)
9. Fontana, T., Vigna, S., Zacchiroli, S.: WebGraph: the next generation (is in Rust). In: Companion Proceedings of the ACM Web Conference 2024, pp. 686–689. ACM, New York, NY, USA (2024)
10. Golomb, S.W.: Run-length encodings. IEEE Trans. Inform. Theory **IT-12**, 399–401 (1966)
11. Kwak, H., Lee, C., Park, H., Moon, S.: What is Twitter, a social network or a news media? WWW 2010, ACM (2010)
12. Rice, R.F.: Some practical universal noiseless coding techniques. JPL Publication 79–22, Jet Propulsion Laboratory Pasadena, California (1979)
13. Versari, L., Comsa, I.M., Conte, A., Grossi, R.: Zuckerli: a new compressed representation for graphs. IEEE Access **8**, 219233–219243 (2020)

Author Index